COST ESTIMATING MANUAL FOR WATER TREATMENT FACILITIES

COST ESTIMATING MANUAL FOR WATER TREATMENT FACILITIES

William McGivney
Susumu Kawamura

WILEY

John Wiley & Sons, Inc.

Copyright © 2008 by John Wiley & Sons, Inc. All rights reserved

Published by John Wiley & Sons, Inc., Hoboken, New Jersey
Published simultaneously in Canada

For general information about our other products and services, please contact our Customer Care Department within the United States at (800) 762-2974, outside the United States at (317) 572-3993 or fax (317) 572-4002.

Wiley also publishes its books in a variety of electronic formats. Some content that appears in print may not be available in electronic books. For more information about Wiley products, visit our web site at www.wiley.com.

Library of Congress Cataloging-in-Publication Data:

McGivney, William.
  Cost estimating manual for water treatment facilities/
William McGivney and Susumu Kawamura.
     p. cm.
  Includes bibliographical references and index.
  ISBN-13: 978-0-471-72997-6 (cloth/CD: alk. paper)
  ISBN-10: 0-471-72997-3 (cloth/CD: alk. paper)
  1.   Water treatment plants–Design and construction–
Estimates–Handbooks, manuals, etc.   I. Kawamura, Susumu.   II. Title.
  TD434.M38 2008
  628.1'60681–dc22   2008003737

10 9 8 7 6 5 4 3 2 1

Contents

Preface

For many years my coauthor, Susumu Kawamura, PhD, and I have been good friends and colleagues in the business of designing and estimating the costs of water treatment plants, along with reservoirs, pipelines, and pumping stations. Susumu performs detailed designing with great precision. My part has been the development of construction cost estimates for the plants and facilities that civil engineers design with great precision. By this I mean that Susumu is paid to be precisely correct in designing a plant, while I am paid to estimate the future cost of transferring the design to a physical plant. In short, he has to be right, and I just have to be close. Of course, I mean this in the best possible way. Estimating is more of an art than a science.

The design criteria and process selection for the complete plants is closely aligned to the Susumu's book, in its second edition *Integrated Design and Operation of Water Treatment Facilities* (Wiley, 2000). We have talked about writing this manual for a number of years, and although it has taken over twice as long as we planned, it passes the estimator's primary test of being close enough for government work.

This manual is specifically for the estimating of construction costs for water treatment plants at the preliminary design level. In order to provide you with a manual that you can use with some confidence, we have compiled the results of our experience and that of many others into a fairly large database of construction costs for separate water treatment processes. The actual historic costs were analyzed, massaged, and separated into component parts representing constructed elements. These primary constructed elements include: civil site work, structures, architectural, process equipment, mechanical piping and valves, electrical, and instrumentation. And each treatment process has more or less the same relative percentage of these elements over at least one order of magnitude of costs, say between $1.0 million and $10.0 million.

Once the cost of the treatment processes is established, they can easily be combined into a complete water treatment plant cost estimate. The manual identifies 43 individual processes or facilities for a "normal" water treatment plant. We have selected nine types of water treatment plants and augmented them with five additional types of advanced water treatment plants. The 43 treatment processes are listed in order for each type of treatment plant along with the number of unit modules and the quantity for the process parameter. The equation form is used to calculate the dollar cost of the complete process, which is summed to a subtotal to which certain mark-ups or allowances are then added to estimate the total direct cost of construction for the plant. To estimate the total raw capital cost, additional allowances are then added for the non-construction or "soft costs." Sample tables for each type of plant are included for 10 mgd and 100 mgd product water flow rates.

The secondary data is adjusted using construction cost indexes bringing them to September 2007 in Los Angeles, California. We used the *Engineering News Record (ENR) Construction Cost Index* which is published and distributed monthly by McGraw-Hill. This index is annotated on each cost curve in this manual (ENR-CCI = 8889).

We have also compiled operations and maintenance costs for these same plants over the same range of product water flow. The O&M costs are similarly set for the current period of September 2007 for labor, power, and chemical usage, as well as maintenance of the facility at an average cost per year.

This manual comes with the complete electronic files in Microsoft Excel on a single compact disk (CD). There are instructions on the CD in the use of the tables for estimating treatment process and total plant costs. The primary data used to generate the curves and establish equations for calculating the costs are not is not part of the manual or CD. The files are not protected nor are they warranted free of error.

We believe this manual will provide you with the basis to estimate construction costs at the preliminary design level for water treatment processes and complete plants. If you compile actual construction costs within your own experience and wish to share them with the authors, we will add them to the database and include the results in the second edition. All confidences will be preserved as they are in regard to the data in this manual. The authors have committed to a second manual for wastewater treatment. So if you find this manual useful, let us know.

—Susumu Kawamura and William McGivney,
Christmas 2007

List of Illustrations

Chapter 1

Introduction to Construction Cost Estimating

1.1 COST ESTIMATING – ART OR SCIENCE?

Is cost estimating an art or a science? My usual response to this question is that cost estimating is both, but more art than science. The science part is made up of engineering and statistics. And the art of estimating is based more on economics and subjective modeling based, and relying on the estimator's experience and knowledge of construction.

Good accurate cost estimating has been the mainstay of human development for at least 8,000 years. Every great empire sustained growth and development because they could afford it. And they could afford this economic development because, among other things, they were good at estimating costs in advance of expenditures. Many other groups suffered through a trial-and-error method of achieving sustained economic development for myriad reasons. But one reason could be that they were very poor cost estimators. This may be a gross oversimplification, but good cost estimating is better than bad.

1.2 STRUCTURE OF THE MANUAL

This manual is not meant to be a rigorous economic analysis or scholarly investigation. It is an outline for preparing good cost estimates for water treatment plants. In this manual the reader will find; basic water treatment plant design philosophy and process schematics; predesign cost estimating methods and procedures; process parameters and their cost curves; and total plant costs, including tables and equation functions; as well as capital and operations and maintenance cost for each type of complete water treatment plant.

The estimating methodology is an amalgam of the best practices of cost estimating and the personal experience of the authors. We have freely used studies and public documents provided by governments and our own historical project data. These tools have provided us with a sound way of developing cost estimates based on specific parameters for individual processes of conventional as well as advanced water treatment plants.

1.3 RULES OF THUMB FOR GOOD ESTIMATES

In the busy life of an engineer or manager, there is rarely enough time to develop a comprehensive detailed cost estimate. So, one may look around for someone not so busy and free to take on the assignment. They may give the assignment to the newest addition to their staff. If this person has enough experience, the estimate will be good. If they have little experience, the estimate will be very poor. And there will be negative repercussions to the budget for design and construction. Cost overruns will run amok, reputations will suffer, and the owner will be very unhappy.

- So, the first rule of thumb is to assign the cost estimating to the best-qualified staff person and give them a copy of this manual to help guide them through the effort.
- The second rule of thumb is to resist the temptation to assume that cost estimates have the same precision as engineering tasks. If they did, they would not be called estimates. Many predesign estimates are carried out to the dollar and much is made of expected accuracy. At this level of estimate a line item estimated to the nearest $10,000 is a reasonable level of accuracy.
- The third rule of thumb is complete the design philosophy and design parameters before estimating the costs. Time is better spent developing solid design parameters such as the detention time, volume of process vessels, and redundancy of process units that will make operation and maintenance of the plant possible.
- Rule of thumb number four is to assign an experienced person to review the estimate. This is obvious on the face of it, but the estimate is usually the last thing to develop when the design budget is exhausted and there is only one day before the report is due.
- The fifth and final rule of thumb is to check the math.

1.4 USE OF HISTORIC DATA

All cost estimators and many engineers and managers keep historic costs in their lower-left desk drawer. This information is a gold mine to their staff and organization. Once adjusted for appropriate indices and level of detail, this data could be added to the cost curves in this book and used to improve the in-house capabilities of the estimator, engineer, manager, and organization. The tables and cost curves contain the formula used to get the best fit for the cost data behind the curves in this manual.

If your experience is the same as the authors, you will find that there appears to be a great variation in data and results. There are many root causes for these variations. Some are the results statistical anomalies; others, economic disparities; yet others, poor record keeping and adjustments. Even with the original, "primary" data, we found coefficients of colinearity (r-squared) in the neighborhood of 0.60 for the total cost of a conventional water treatment plant. And got r-squared(s) of 0.35 for pumping stations. In summary, do not expect precision, but constantly test your assumptions, recheck the math, and review the work of others.

1.5 ADJUSTING THE NUMBERS

Historic costs have a way of remaining constant. They represent the actual price of goods and services at some time in the past. They can be adjusted to another time or place on the basis of a cost index published by either the government or a private entity that is generally accepted by the industry or constituency it represents. It is important that the estimator select the most reliable index and apply that index to the historic cost to compare it to other costs, either actual or estimated. Once adjusted, the resulting cost is no longer considered primary data.

Adjusting actual costs from some time in the past to the current period presumes that the goods and services that made up historic cost have not changed and the costs for all components have changed in exactly the same way. Making this adjustment can introduce inaccuracies into the estimate. Adjusting the actual cost from place to place either across the country or from country to country is even riskier. And making both types of adjustments can eliminate any reasonable expectation of accuracy.

Our recommendation is to make at least three separate estimates of the cost using different means and assumptions. The cost curves and tables in this manual are one way to go about the estimating process. Getting input from someone of greater experience is the second. And using actual costs from published documents as comparisons could be the third. In this way the estimator is able to plot a triangle of points and test the individual process or complete treatment plant cost model for reasonableness.

Chapter 2

Water Treatment Processes

2.1 BASIC PLANT DESIGN PHILOSOPHY

Construction cost estimating at the preliminary design phase of a project is dependent on the basic design scheme, including sketches of the project. A properly and clearly prepared design philosophy is essential for the success of the design and construction of all treatment facilities. The well-prepared preliminary design construction cost estimate will form the basis of an accurate capital projects budget. This type of cost estimate is based on experience and intuition rather than the more rigorous detailed engineer's estimate.

Following a half-century of water and wastewater treatment design, construction and plant operational experience a pattern of successful design development has become clear. There are ten basic rules or commandments for a successful design project.

The Ten Commandments for design project are as follows:

1. You shall make a careful analysis and evaluation of the quality of both raw and required finished waters.
2. You shall undertake a through evaluation of local conditions.
3. The treatment system developed shall be simple, reliable, effective, and consist of proven treatment processes.
4. The system considered shall be reasonably conservative and cost-effective.
5. You shall apply the best knowledge and skill available for the design.
6. The system shall be easy to build and constructible within a reasonable length of time.

5

7. The system shall be easy to operate with maximum operational flexibility and with minimum operation and maintenance costs.
8. The facilities shall be aesthetically pleasing with no adverse effect on the environment.
9. Design engineers shall perform services only in the area of their competence. Get help from qualified experts in areas outside your expertise.
10. You shall respect and owner's knowledge and experience and incorporate his wish list of additional features if they are within the established budget.

2.2 BRIEF DESCRIPTION OF BASIC WATER TREATMENT

Early water treatment systems were simple batch operations designed for individual households. These processes included boiling, simple filtration, and coagulation and filtration utilizing naturally available inorganic or organic coagulants. However, from the seventeenth century onward, it was necessary to create facilities capable of treating large quantities of water to supply larger human settlements. The treatment of water based on scientific principles began in Europe around the mid-1800s. During this time, water treatment professionals in England undertook the elimination of water-borne diseases such as typhoid and cholera.

The application of chlorine to potable water supply systems in England, during the 1850s, followed the scientific validation of germ theory. However, it soon became evident that chlorination was ineffective when applied to cloudy water. This gave rise to the process of slow sand filtration (0.05 gpm/sf or 0.125 m/hr filter rate), which removed suspended solids before the application of chlorine. This first era of water treatment was control of pathogenic bacteria by chlorination preceded by slow sand filtration.

During the late nineteenth century, the Louisville Water Company in Kentucky began pretreating raw water with alum coagulation followed by clarification and the use of rapid sand filters (2 gpm/sf or 5 m/hr filter rate). This new process was urgently needed. A significant increase in population and rapid industrial growth placed a demand on the water system that the slow sand filters could not meet. This development was the beginning of the water treatment plants of today.

Drinking water quality standards were relaxed until the middle of the twentieth century. Only minor changes to the basic conventional treatment processes occurred until the late 1960s. The object of the water treatment in this period was to produce sufficient amount of water "safe" from pathogenic bacteria. Water treatment engineers, from late 1960s to 1970s, concentrated their effort on designing the lowest-cost treatment system to produce "safe" drinking water. High rate filtration and high hydraulic loading for a sedimentation basin with tube settler or plate settler modules and the use of ozone as an advanced treatment process have become popular since the mid-1990s.

The beginning of modern water treatment design started after the Second World War. High-technology industries flourished in the postwar years in industrialized nations such as United States. As a result, large quantities of untreated synthetic industrial wastes were discharged into nearby water courses, the oceans, or the atmosphere, or dumped into and onto the land. Consequently, serious global environmental pollution resulted in more stringent drinking water quality standards, and new advanced treatment processes were urgently needed.

During the early 1970s, the Environmental Protection Agency (EPA) was established and the Safe Drinking Water Act (1974) and its amendment (1986), subsequently passed by the U.S. Congress, set stringent drinking water quality standards.

The motto of water treatment had now become "make large quantities of 'good' quality water."

The issues after mid-1990s are control of protozoa, especially *Cryptosporidium* and *Giradia*; control of disinfection process byproducts as well as arsenic; disposal of treatment residues; and the supplying of noncorrosive water. Recent treatment issues coming up are treatment of xenobiotics, which are related to small amount of pharmaceutical and drug residuals in source of waters, as well as control of taste and odor.

Today, we have advanced water treatment technology and thousands of miles of water distribution systems. However, the field of water treatment faces new problems such as a limited source (less than 3% of water on earth) of easily treatable water for potable water, heavy industrial and human activities, and the population explosion.

The project development and project delivery procedure in recent years have been shifting away from traditional ways. The old way was having a single group of civil engineers handle the majority of design work, supported by mechanical, electrical, and architectural engineers.

However, the regulatory requirements and complexity of the projects now require a multidisiplinary design team.

The traditional designing of water treatment plants includes a professional engineering firm or owner's in-house staff who prepares specifications and drawings. Sealed bids are received and contractor(s) selected based on the lowest responsible bidder. The design team performs construction management services until the facility is completed and commissioned. After commissioning by the design team and the owner, they begin operating the facilities.

In early 1990s, changes were taking place in traditional project delivery. The idea was the incorporation of design, construction, and operation of the facilities with a new financial/political arrangement called "privatization." Privatization as its name implies is turning over all or part of the facility development and operation to a privately held entity. These schemes include; design-build-operate (DBO), design build-maintain (DBM), public-private-partnerships (PPPs), and long-term contract operation.

The recent popularity of privatization for domestic water utilities is the result of internal and external competition. Contributing factors include increased regulatory requirements for upgrading existing as well as new plants, negative consequences from different levels of maintenance, public resistance to rate increases, and the financial crisis faced by many public utilities. However, privatization projects also have negative aspects, including less than optimum safety as well as reliability for plant and a tendency toward operational inflexibility. These negative issues are mainly due to attempts to improve profitability, reduce costs by rapid facility construction, and keep operational costs at a minimum. This is also true for wastewater treatment facilities.

2.3 BASIC CONVENTIONAL WATER TREATMENT PROCESSES

Figure 2.3 above shows the relative size and layout of the treatment processes of a conventional water treatment plant. The basic conventional treatment train for surface water treatment consists of coagulation with rapid mixing followed by flocculation, sedimentation, granular media filtration with final disinfection by chlorine. This treatment process train is a standard requirement for municipal water treatment by the Department of Health Services (DHS) of each state as well as the Ten State Standards, which apply to the ten states in the Midwest Region and the

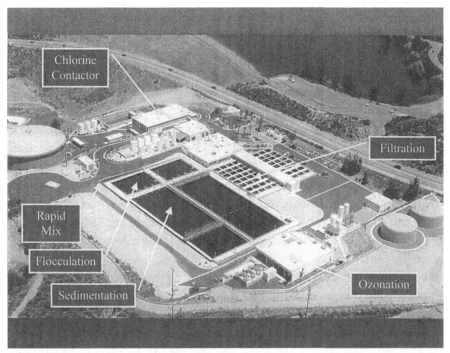

Figure 2.3   Overhead View of Water Treatment Plant

East Coast Region of United States. However, the basic treatment trains can be modified, dependent upon the quality of raw water and the finished water quality requirements.

For instance, where the raw water quality is good, sedimentation process can be excluded from the basic treatment train. This process system is Direct Filtration. In some instances, both regular flocculation and sedimentation process are replaced with coarse media flocculation/roughing filter process in front of regular granular media filtration. In other cases, the filtration process is preceded by flash mixing of a coagulant. This is the In-Line Filtration or Contact Filtration process. However, these modified conventional treatment processes must have a variance permit from the governing regulatory agencies before design and facility construction.

If surface waters have high levels of turbidity, hardness, total organic carbon (TOC), microorganisms including algae, taste and odor, and other unwanted substances, then certain additional process or modifications of the conventional process and plant operation will be necessary. Flash mixing of coagulant at the head of plant is essential and the water-jet diffusion type is the most effective system. The current flocculation basin is a rectangular basin and vertical shaft mechanical flocculators with hydrofoil type mixing blades. An earlier design included a horizontal shaft with paddle type mixing wheel. However, an improved baffled channel design (helicoidal flow pattern) is currently in use. Common sedimentation tanks are rectangular horizontal flow type with or without high rate settler modules such as tube settler or plate settler modules. A mechanical sludge collection system is a part of the sedimentation system. A few proprietary units use a combination of flocculation and clarification processes.

The common filtration system consists of gravity filters with granular media beds. The anthracite and sand dual-media bed has been a standard filter bed since the 1980s. Surface wash systems for 6″ to 18″ depth of bed depending on the system used, as well as air scouring wash systems that scour the entire filter bed, with a backwash and filter-to-waste provision have become common. The clearwell usually provides at least 4 hours of finish water storage capacity. The clearwell should be baffled to minimize flow short-circuiting, and it must be covered.

Chemical storage and feed system are an important part of the treatment plant. The sludge handling and disposal is an essential facility of water treatment plant. These items are discussed later in this chapter. A few water treatment plants require intermediate pumping. Intermediate pumping facilities can become expensive when required by hydraulic analysis. Plant security systems are critical facilities due to the potential for acts of terrorism.

Basic ground water treatment uses granular media filtration process followed by chlorination. If the water quality of the source is exceptionally good, only disinfection by chlorine may be required. However, an oxidation process may be needed when high levels of soluble iron, manganese, and other substances exist in the source water.

The granular filtration process is always included in the basic treatment process because it is the main barrier to keep suspended matter, including microorganisms, from passing into the potable water supply. Over the last forty years, filter design has become either dual-media bed or coarse media deep bed with or without a thin fine sand layer at the bottom. The filtration rate for these filters is usually limited to 6 gpm/sf (15 m/h) by regulatory agencies. However, several water treatment plants on the West Coast are achieving a flow rate of 8 to13 gpm/sf (20 to 32.5 m/h) with pre-ozonation under variance permits issued by the California Department of Health Services. Figures 2.3.1a, 2.3.1b, and 2.3.1c

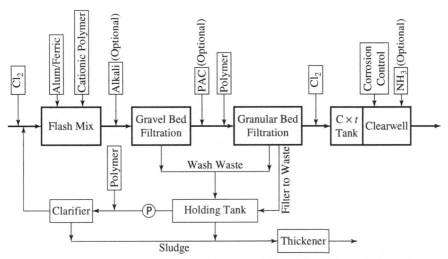

Figure 2.3.1a   Two-Stage Filtration Process

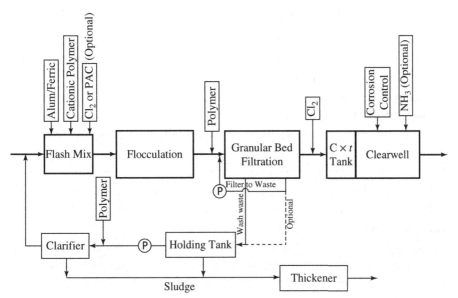

Figure 2.3.1b   Direct Filtration Process

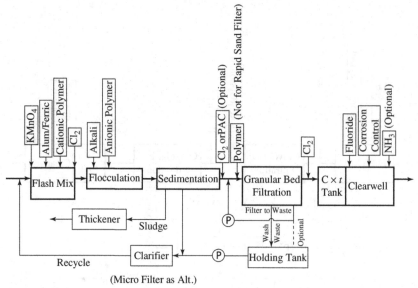

Figure 2.3.1c Conventional Treatment Process

show three diagrams of a basic conventional treatment processes with slightly different chemical application systems.

2.4 ADVANCED WATER TREATMENT PROCESSES

As described earlier the EPA has promulgated the Surface Water Treatment Rule (1989) and the Interim Enhanced Surface Water Treatment Rule (1998) in order to provide not only *safe* but also the *best quality* drinking water for the public. The major elements of these rules include removal of total organic carbon (TOC) from raw water to certain targeted levels in order to control the disinfection byproducts (DBPs) and inactivation or removal of *Cryptosporidium oocysts*, which regular chlorination cannot achieve. There are many other Maximum Contaminant Levels (MCL) for drinking water quality standards for inorganic and organic chemicals, microbiological contaminants, disinfectants, radionuclides, turbidity, and other conditions.

Since the basic conventional water treatment processes cannot achieve these requirements unless the source of water is exceptionally good, several new treatment process technologies have been developed and implemented in recent years.

Ozonation, granular activated carbon adsorption, high-speed micro-sand settling process, high-rate dissolved air flotation (DAF) process, magnetic exchange (MIEX) process, new type of UV disinfection process, and advanced membrane filtration process (MF, UF, NF, and RO) are considered as major advanced water treatment processes of in late twentieth century to early twenty-first century. These new treatment processes are used in conjunction with the basic conventional treatment process described earlier.

Figures 2.4.1a, 2.4.1b, and 2.4.1c illustrate three examples of advanced water treatment plants.

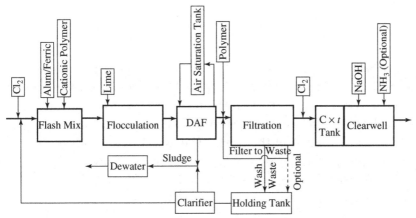

Figure 2.4.1a   Dissolved Air Floatation (DAF) as Pretreatment Process

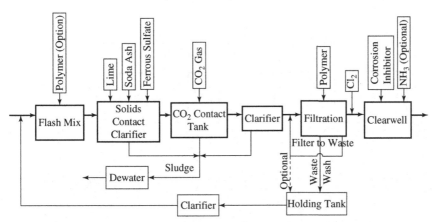

Figure 2.4.1b   Lime and Soda Ash Water Softening Process

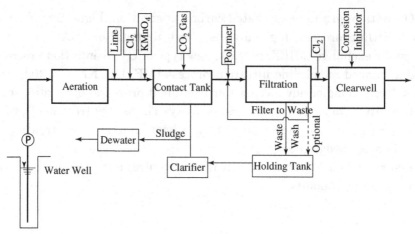

Figure 2.4.1c   Typical Iron and Manganese Removal Process

These advanced treatment processes are also being incorporated into wastewater treatment design as advanced treatment processes for water reuse purposes. Desalination and water reuse are growing water treatment technologies because of a growing shortage or contamination of raw water in many regions of the world. The as yet unknown consequences resulting from global warming, whatever the cause, may rapidly increase the need for water reuse. Figures 2.4.2a, 2.4.2b, and 2.4.2c are examples of additional, advanced water treatment plant design.

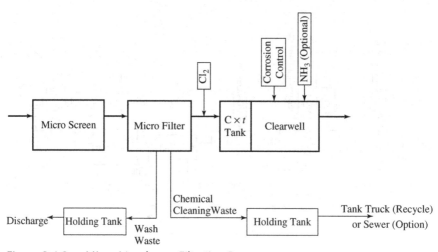

Figure 2.4.2a   Micro Membrane Filtration Process

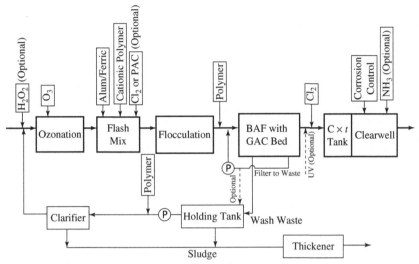

Figure 2.4.2b Direct Filtration Process with Pre-Ozonation

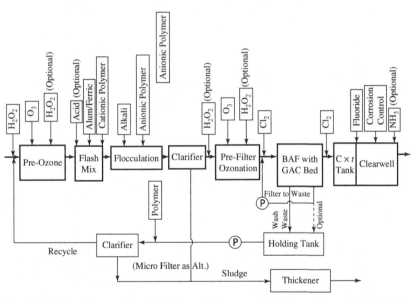

Figure 2.4.2c Conventional Treatment Process with Ozonation and GAC Filters

In the following chapters, preliminary construction costs will be developed for each of these nine scenarios, with a design plant flow of 10 MGD and 100 MGD. The data used for the cost curves was collected over many years from multiple sources.

Chapter 3

Solids Handling and Disposal

3.1 SOLIDS HANDLING

Solids handling begins with thickening the collected sludge in order to increase the solids content from 0.5% produced in the clarifiers to anywhere from 3% to as much as 70% solids. The sludge removal process will waste less than 0.5% of the plant flow unless the sludge is dewatered and the liquid returned to the head of the plant. This thickening is aided by the addition of a polymer to the collected sludge increasing the percentage of solids in the sludge liquor. The sludge is then processed either through gravity thickening or by mechanical thickening, increasing the percentage of solids. The minimum percentage of solids allowed at most disposal sites is dictated by a combination of federal and local regulations.

3.2 SLUDGE THICKENING

There are several means to handle this task, including sludge lagoons, gravity thickening tanks, and dissolved air flotation tanks. Each process will produce a different percentage of solids and have different construction and operations costs. The construction costs of these processes are among those detailed in Chapter 5 of this manual.

It may be possible to transfer the cost of solids handling simply by adding the sludge to the local sewer connection. The local sewage disposal utility can provide the fee structure to be used along with a meter attached to the sewer lateral that will keep track of the discharge. For a small water treatment plant this transfer fee will usually be lower than the capital investment and O&M costs of a solids facility. Some existing water treatment plants utility use existing sludge lagoons where the liquid seeps into the ground leaving the solids at the surface. Lagoons are by far the cheapest from both a construction and operating cost. If there is sufficient land available and this type of system is permitted, it is a very cost-effective solution.

To avoid compromising the ground water aquifers the filtrate must be captured and not allowed to seep into the ground. Sand drying beds are designed with liners, drains, and sumps to catch the filtrate. They also provide a higher percentage of solids and allow the capture of the filtrate to be disposed of in the sewer.

Gravity thickeners and dissolved air flotation tank construction costs are necessary when land is scarce, particularly when an existing plant is going through phased expansion. These processes are commonly used as the first stage of solids handling at water treatment plants.

3.3 SLUDGE DEWATERING AND DRYING

Once the solids pass through a thickening process they are further processed to remove additional water and increase the percentage of solids by volume. With the exception of sand drying beds all other drying methods require mechanical means and significant investment. Mechanical equipment for dewatering and drying include; solid-bowl centrifuges, belt-filter presses, recessed plate filter presses, vacuum filters, rotary sludge dryer, and incinerator. Centrifuges and filter presses are commonly used and are among the processes identified in Chapter 5. Vacuum filters and rotary sludge dryers with incinerators are not addressed here.

Chapter 4

Construction Cost Estimating at Predesign

4.1 CONSTRUCTION COST ESTIMATING

A basic element of the design process is facilities construction cost estimating. Establishing a reasonable construction budget during the predesign work will add direction and integrity to the design process. The predesign will typically consider multiple process alternatives. Each type of process will have a different construction cost. The owner must take into account interest rates, administrative/legal costs, design engineering costs, land use, and local political considerations. Invariably the issue of design engineering and engineering support during construction will be carefully reviewed and negotiated in part on the integrity of the predesign process. Each of the process alternatives will also have unique operation and maintenance costs dependent on the requirement for labor, energy, chemical and other consumables.

Since design development is a dynamic process, it is very important that the estimated cost of the project be periodically checked against the capital budget. Therefore, a series of cost estimates are prepared and compared to the previous ones. These estimates should be as detailed as possible, based upon the increasing level of information available. It is recommended that written guidelines for each of type of construction cost be developed. An initial set of guidelines is offered below.

The Association for the Advancement of Cost Engineering, (AACEi) has established a comprehensive set of standards and guidelines for this purpose. It is imperative that the entire design team, and all project stakeholders, be kept fully informed, and be held responsible for their participation in the establishment of budgets for design, and ultimately for the final construction cost of every project. A project-centered approach coupled with an effective, open channel of communication will prevent unfortunate surprises to escalating or uncontrolled costs. It will further help ensure a professional engineering environment, directed at problem solving, rather than one deteriorating into reactionary, or adversarial, relationships between members of the design team.

More specifically, these guidelines establish the criteria, format, usage, accuracy, and limitations of the various types of construction cost estimates. To accomplish this task it is necessary to implement the following criteria:

- Define the type of cost estimates and a detailed narrative called the Basis of Estimate to be prepared during the various phases of a projects development. Within these definitions would be the expected accuracy of the cost estimate as well as limitations on its value and use.
- Define the responsibility for the preparation and review of these cost estimates.
- Define the cost estimating method relative to the construction and capital improvement costs of the facilities.
- Develop the procedures to prepare and review the cost estimates on a uniform basis.

Once developed, these standards can be implemented on all design projects and capital improvement programs. A consistently applied set of guidelines and growing project database as well as local unit pricing data can provide a more accurate, less problematic cost estimate.

4.2 CLASSES AND TYPES OF COST ESTIMATES

Construction cost estimates are categorized into five classes: 5, 4, 3, 2, and 1, in reverse numerical order by level of detail available depending on their use. Each cost estimate should include a "Basis of the Estimate" narrative as discussed below. As the level of detail required in performing the cost estimate increases, the labor and experience of the estimating staff required to complete the material take-off and pricing rises significantly. The estimating accuracy discussed below, pertinent to each class of cost estimate, is not meant to represent absolute limits or guarantees, but instead to establish a most likely range within which the final project construction cost will fall.

4.3 PREDESIGN CONSTRUCTION COST ESTIMATING

The first of these is the *predesign cost estimate*. The intent of this manual, given the lack of design detail, is to assist in the establishment a realistic estimate of the cost and time components, based on a combination of unit costs and process parameters. This cost estimate is typically defined as a Class 5 cost estimate with an expected accuracy of $+50\%$ to -30% of the average bid price for construction. This type of construction cost estimate is generally used for the development of capital improvement plans, master plans, and feasibility studies. Predesign construction costs are useful in the comparisons between project cost and between specific process alternative costs. As a result, the predesign cost estimate is particularly sensitive to assumptions and qualifications.

This is most important when designing for plant rehabilitation and operating facility expansion. It is also significant when comparing process types and system components. And finally, it can be used when comparing costs for alternatives that include ultimate capacity versus current flow rates.

4.4 DEFINITION OF TERMS

Accuracy of the Estimate

The accuracy of a predesign cost estimate is taken from the guidelines of the American Association of Cost Engineers, International (AACEi) as a percentage range for estimating purposes. The accuracy ranges are identified by the use of percentages $+/-$ in reference to the expected actual construction cost of the work. These ranges differ with the type of estimate performed. (i.e., the Class 5 predesign cost estimate can be expected to range from $+50\%$ to -30% of the actual cost of the project.) A graphic representation in the appendix identifies the accuracy range for each class and type of cost estimate.

Allowance for Additional Direct Costs

Due to the preliminary stage of the project, some conditions affecting the pricing and productivity could change. This percentage allowance is intended to cover work items not yet quantified but known to exist in projects of this type and size. This allowance is a variable percentage (%) of Total Direct Cost.

Construction Costs

Construction costs are the sum of all individual items submitted in the successful contractor's winning bid through progress of the work, culminating in the completed project, including change order costs.

Construction Cost Trending

A construction cost "trending" is the preparation and updating of the project construction cost estimate over time. As the design process continues, the project becomes more defined and as more detailed engineering data becomes available, "trending" provides a basis for the analysis of the effects of these changes. These design and construction-related issues are the documented, logged, and analyzed on a regular periodic basis during the design phase of the project. As each issue is resolved, it is included in the "trended" cost estimate.

Contingencies

Contingencies are defined as specific provisions for unforeseeable cost elements within the defined project scope. This is important where previous experience relating estimates and actual costs has established that unforeseeable events are likely to occur. Allowances for contingencies are an integral part of the estimating process. Contingency analysis of cost estimates is a useful aid to successful project performance. The periodic review and analysis of these contingencies provide a myriad of opportunities for project management to assess the likelihood of overrunning a specified dollar amount, budgetary limitations, or time commitments. However, many owners are hesitant to include unidentified contingencies in their budgets. This may be because the owner believes that the engineer can and should identify all issues that impact the cost of construction in advance of completing the design process. This is a fair assumption, but it invites "cost creep" during the design process and forces the owner to increase the construction budget of the project or lose precious time in redesign.

Cost Indexes

Cost indexes are a measure of the average change in price levels over time, for a fixed market basket of goods and services. Commonly used indexes effecting construction costs are:

- *The Consumer Price Index (CPI)*: The Bureau of Labor Statistics of the U.S. Department of Labor produces monthly data on changes in the prices paid by urban consumers for a representative basket of goods and services. It is applicable, in a general sense, to the monthly and annual change in the cost of goods and services in the end user market. These would be price- level changes in aggregate including construction related goods and services for labor and incidental materials.
- *The Producer Price Index (PPI)*: This is also prepared by the U.S. Bureau of Labor Statistics and measures the average change over time in the selling prices received by domestic producers for their output to the wholesale market. These are price-level changes in specific manufactured products, including construction related goods like cement, steel, and pipe.
- *The Engineering News-Record Construction Cost Index (ENR-CCI)*: This is applicable in a general sense to the construction costs. Created in 1931 the ENR-CCI most nearly tracks the price-level changes in major civil engineering capital construction costs over time.
- *The Handy-Whitman Index of Water Utility Construction Costs*: This has been maintained since 1949 and is applicable to water and wastewater treatment plants. This index more closely approximates aggregate price-level changes in more complex treatment plants and pumping stations.
- *The Marshall and Swift National Average Equipment Cost Index*: This is specifically for the tracking of complex equipment price-level changes. It is useful in the indexing of pure equipment costs over time.

Escalation

All cost estimates are prepared using current dollars. In preparing a cost estimate, an evaluation may be made to determine what effect inflation may have on the cost of the project extending some time into the future. An escalation factor is then applied to the total cost or "bottom line" of the estimate. A published cost index is the basis used for escalation should be localized, reputable, and reflective of the construction industry. More specifically, the cost index used must be particular to the type of project being designed and constructed. This escalation may be used in conjunction with the cost-of-funds to prepare a present worth analysis of various project alternatives or construction phasing requirements.

4.5 ESTIMATING METHODOLOGY

The cost-estimating guidelines identified here comply with conventional, professional standards established by the AACEi. Project and process parameters must be developed and a preliminary site layout established. Process parameters for this manual are in English units (i.e., gallons, feet, tons, etc.). The data used for this manual was compiled from actual plant and process construction in the United States since 1970. The cost data has been "normalized" to a current construction cost index published monthly by the *Engineering News Record*, McGraw Hill, as the Construction Cost Index (CCI).

Cost Capacity Curves

Cost capacity curves are charts that describe average costs of an item as a function of capacity. The typical cost curve for water treatment processes is a smooth line drawn or fitted through a scatter of real data points adjusted to a fixed time by which the data is "normalized." The information on that chart would define the construction cost for a process in dollars at various plant flow capacities at a fixed point in time.

The cost and cost capacity curves relate to various process treatment alternatives, pumping station capacities, reservoir storage capacities, and pipeline sizes and types, as well as other civil engineering projects. Other sources of cost and cost capacity curves cited below may also prove effective and reliable when used correctly. Some of these published documents are no longer in print, but copies may be available at various university or government locations.

- "Innovative and Alternative Technology Assessment Manual." February 1980 EPA/430/9-78-009 MCD-53.
- "Operation and Maintenance Costs for Municipal Wastewater Facilities." September 1981 EPA/430/9-81-004 FRD-22.
- "Construction Costs for Municipal Wastewater Treatment Plants: 1973 – 1978." April 1980 EPA/430/9-80-003 FRD-11.
- "Treatability Manual, Volume VI, Cost Estimating." July 1980 EPA/600/8-800-042d.

- "Estimating Costs for Water Treatment as a Function of Size and Treatment Plant Efficiency." August 1978 EPA/600/2-78-182.

Basis of Estimate

A brief description, in narrative form, of the work scope, assumptions, and qualifications with details particular to the project should be presented with the finished cost estimate.

Structure of the Estimate

The estimate is organized by both alternative treatment trains and specific processes. A treatment train will include all processes specific to that alternative. If there are alternative processes under consideration, a separate treatment train must be created to include all necessary processes associated with that alternative.

Once the alternative treatment trains are identified and their parameters calculated, their individual construction costs can be calculated from the process graphs or equations in this manual. Once the alternative process treatment trains are identified and their construction costs estimated, additional site-specific parameters may be applied.

These additional cost parameters could include: interconnecting conduit and yard piping, site demolition, earthwork, paving and grading, landscaping and irrigation, and electrical and instrumentation infrastructure. Separate curves and their parameters for these additional costs are included in this manual.

Estimate Global Mark-Ups

These mark-ups typically pertain to specific allowances for Escalation, Contingency, Construction Management, Inspection & Construction Administration, Design, Administration & Legal, Rights of Way Acquisition, Environmental Mitigation, and Permitting. The construction cost estimate may include any or all of these, depending upon the requirements set by the client.

Comparison of Alternative Process Construction Costs

When developing the alternatives, it is best to estimate costs for the entire treatment train that includes the selected alternatives. In this way, a complete picture can be developed for the alternative analysis. Process parameter values and their relative cost of construction may not differ much. But, depending on how the selected processes are laid out and connected on the site, the total cost of the train including the alternative process can be dramatically different. Land constraints, process hydraulic requirements, and subsurface conditions can dramatically increase the overall cost of the facility many times the difference in cost of two or more alternatives.

4.6 CAPITAL IMPROVEMENT COSTS

A capital improvement program can be made up of a number of individual projects and span many years of development, design, and construction. The cost of a capital improvement program is usually developed without significant design input. Realistic budgets established early in the predesign phase can improve the likelihood of the program's success. Knowing the reasonable cost of design, construction, and operation disruption on an annual basis can provide labor and cost savings and avoid delays that will invariably drive costs much higher than expected. A good realistic plan, even one based on parametric ratios, can provide tools for a more successful outcome or reduce the likelihood of an embarrassing failure.

Starting with a set of realistic construction cost estimates for multiple treatment trains that include unique costs for construction, operation, and maintenance, and additional nonprocess costs assists in evaluating and selecting the project(s). By including the estimated time for the development, design, and construction the project costs can be spread over an annual calendar to assess the availability of capital, offsetting revenue, and staffing required to make the program a success. And the estimate of cost and time begins with a reasonable and realistic process construction cost. As the estimating process is developed in Chapter 5, an allocation of costs for each parameter group will be made and a total capital improvement cost model will be presented as an example of how it all fits together.

Regulatory Impact

As was shown in Chapter 2, the historic changes in water treatment regulations by both state and federal government agencies has, over the preceding century, accelerated as a result of public concern, scientific testing, and technological improvements. With the increased burden of a growing population, limitations on waste disposal, and resistance to change in the acceptance of water reuse, we can expect the acceleration in regulatory response to continue. As a result, many of the treatment processes that are the basis of water treatment design could become outmoded as more restrictive regulations on water quality are imposed. The present cost of design and construction of advanced treatment facilities for microfiltration and desalination are included in this manual.

Operations and Maintenance Costs

The costs of operating and maintaining an existing or new treatment facility can vary from plant to plant even for the same owner. The operating costs are dependent to a great degree on the energy requirement and chemical dosage. These costs are directly related to the quality of the raw water that must be brought up to a minimum "good" quality set by regulatory agencies and plant hydraulics, which are dictated by the plant hydraulic profile. If intermediate pumping is necessary, then the energy costs and maintenance costs for continuous pumping drive O&M cost higher. Labor costs are a factor but can be overshadowed by the cost of energy and chemical consumption. Larger chemical storage and feed facilities are also regulated and becoming expensive and time-consuming to maintain. Simple parameters that are related to these processes are included in this manual. O&M cost curves are for each type of plant are discussed and shown in Chapter 6.

Chapter 5

Water Treatment Predesign Construction Costs

5.1 INTRODUCTION

In this chapter, we will identify and examine the parameters developed for estimating construction costs. These parameters will then be applied to the nine types of water treatment plants specified in Chapter 2, at design flow rates of both 10 and 100 million gallons per day (MGD). The results will be made into tables for 43 different processes equations based on predesign parameters at the two design flow rates. We will also present cost curves for advanced treatment plants including: four types of seawater desalination plants, ultra-filtration and membrane filtration.

In Chapter 2, we identified nine types of water treatment facilities, each characterized by their unique design parameters and processes. These water treatment designs are listed again below by figure and name.

2.3.1a Two-Stage Filtration
2.3.1b Direct Filtration
2.3.1c Conventional Treatment
2.4.1a Dissolved Air Flotation
2.4.1b Lime and Soda Ash Softening
2.4.1c Iron Manganese Removal
2.4.2a Micro Membrane Filtration
2.4.2b Direct Filtration w/Pre-Ozonation
2.4.2c Conventional Treatment w/Ozonation and GAC Filters

Each of these plants types has unique processes and operating parameters differentiating them from one another by purpose and ultimately by construction costs based on those parameters. The two design flow rates are one order of magnitude apart at 10 and 100 million gallons per day. Historical process costs have been gathered, sorted, tabulated, and graphed to show the relationship between the process parameter and construction cost. Results consistently show it is not a one-to-one relationship. For example, it doesn't cost twice as much to design and construct a circular clarifier with a 100-foot diameter as to one with a 50-foot diameter. We have identified forty-three specific processes or facilities that are currently used in water treatment plant design, prepared cost curves and compiled them into a total plant cost including nonprocess costs common to the construction of these plants.

These cost and cost capacity curves have been developed for specific treatment process alternatives, pumping station capacities, clearwell storage capacities, and process pipeline sizes and types, as well as other components, including engineering design and construction support costs are made a part of this *Cost Estimating Manual*. Other sources of cost and cost capacity curves cited below may also prove effective and reliable when used correctly. Some of these published documents are no longer in print but copies may be available at various university or other sources.

- *Estimating Costs for Water Treatment as a Function of Size and Treatment Plant Efficiency*, August 1978 EPA/600/2-78-182.
- *Innovative and Alternative Technology Assessment Manual*, February 1980 EPA/430/9-78-009 MCD-53.
- *Construction Costs for Municipal Wastewater Treatment Plants: 1973–1978,* April 1980 EPA/430/9-80-003 FRD-11.
- *Treatability Manual, Volume VI, Cost Estimating*, July 1980 EPA/600/8-800-042d.
- *Operation and Maintenance Costs for Municipal Wastewater Facilities*, September 1981 EPA/430/9-81-004 FRD-22.

5.2 TREATMENT PROCESS AND COST ESTIMATING PARAMETERS

The processes we identify here have general parameters such as: square feet, gallons per day, lineal feet, and so on. These parameters are a way of comparing the relative size or function of the process against historic, updated construction costs. In this way, we generate a parametric curve so that we can estimate the construction cost. Included in the appendix is a short table of common conversions to the metric system used in the design of water treatment plants. The curve functions developed here have been calculated using the standard single-variable "trend line" functions available in an electronic worksheet. In most cases, the "trend line" function that best fit the data was a simple equation represented by $(y = aX^n)$. In some cases, the equation that best fit data was of the form $(y = aX + b)$. Both types of equations are supported by economic theory within a parametric range of one order of magnitude.

Since the historic costs have occurred over many different years and places within the United States, they were "normalized" to a common time and place by using the published cost indices identified in Chapter 2. These curves were updated to an ENR CCI = 8889 for Los Angeles, California (April 2007). Since the historic data has already been updated to a common period and location, future updating of the cost curves are made by multiplying the cost by the change in the applicable index, that is, for a 10% increase in the index the cost is multiplied by 110% to get the updated cost.

The following table of forty-three water treatment processes, facilities and additional nonprocess cost multipliers are the result of the analysis of historic actual costs. The resulting Total Project Cost is the sum of all construction costs, mark-ups, and engineering, legal, and construction administration costs in current dollars. This type of predesign cost estimate provides a common basis for evaluating process alternatives against total project costs. By applying these costs curves to phased design and construction, a Present Worth analysis, using interest rates, and operations and maintenance costs can assist the engineer and owner in choosing between alternatives.

The table below lists an array of processes, cost equations, and range of application parameters. These cost equations and source data are compiled into individual cost curves and detailed in the following section.

Before using the cost equations or curves, you must know the design criteria for each process. For example; for the Clarifiers, Process Nos. 23 & 24, Circular Clarifiers (No. 23) must use its accepted hydraulic loading of 1.0 gpm/sf to 1.4 gpm/sf (avg. 1.2 gpm/sf). But Rectangular Clarifiers (No. 24) have a hydraulic loading rate of 0.5 gpm/sf to 1.0 gpm/sf (avg. 0.75 gpm/sf) without plate settlers. If you use 0.75 gpm/sf as the loading rate circular clarifiers the cost becomes very expensive in relation to the rectangular clarifiers.

5.3 COST CURVES

Each of these processes is also represented by a unique cost curve like the one for Figure 5.5.23 Circular Clarifiers. The curve is fit to the data and represented by an equation and an r^2 or measure of colinearity. Since the original data has been consolidated and an average value for each unit of parameter, the r^2 simply tells us if the best fit curve type is "power" $(y = aX^n)$, "polynomial" $y = aX^2 + bX + c$, or "linear" $(y = aX + b)$ is the most appropriate.

Using either the equation function in Table 5.2.1 or scaling off the appropriate process curve, selecting the parameter of 17,000 SF for clarifier floor area drawing a perpendicular line from the curve where x = 17,000 and a second line to the y-axis and estimate the value for the construction cost at approximately $1,4250,000. From the equation function $y = 3470.6x^{0.6173}$ where x = 17,000 square feet of floor area the value for the construction cost is $1,424,736. Although the equation delivers a more precise arithmetic answer, it is no more accurate than a rough line drawn on the graph. The likelihood of either number being correct is the same.

5.4 ESTIMATING PROCESS AND TOTAL FACILITIES COST

Each of the forty-three water treatment processes have a range of application, and we will briefly discuss the limits, physical characteristics, and ultimately the costs as they apply to the nine types of treatment plants for both 10 MGD and 100 MGD.

5.5 INDIVIDUAL TREATMENT PROCESS COST CURVES

5.5.1 Chlorine Storage and Feed from 150-lb to 1-ton Cylinders

Chlorine gas is purchased from the producer and delivered to the site to be used as a disinfectant for the finish water delivered to customers of the water treatment plant.

The smallest application is a 150-lb vertical tank with an eductor feed system. This application might be used as the only treatment for a small system where the raw water is taken from a shallow aquifer and stored above ground for relatively short periods of time.

Table 5.2.1 General Cost Equations for Water Treatment Processes with Parameters, Minimum and Maximum Limits

No.	Process	Cost Equation	Min	Max
1a	Chlorine storage and feed 150# cylinder storage	$\$ = 1,181.9 X^\wedge 0.6711 \quad X = Chlorine\,feed\,cap.\text{-}Lb/d$	10	200
1b	Chlorine storage and feed 1-ton cylinder storage	$\$ = 5,207.41 X^\wedge 0.6621 \quad X = Chlorine\,feed\,cap.\text{-}Lb/d$	200	10,000
2	On-site storage tank with rail delivery	$\$ = 63,640 X^\wedge 0.2600 \quad X = Chlorine\,feed\,cap.\text{-}Lb/d$	1,980	10,000
3	Direct feed from rail car	$\$ = 69,778 X^\wedge 0.2245 \quad X = Chlorine\,feed\,cap.\text{-}Lb/d$	1,980	10,000
4	Ozone Generation	$\$ = 31,015 X^\wedge 0.6475 \quad X = Ozone\,Gener.\,Cap.\text{-}Lb/d$	10	3,500
5	Ozone Contact Chamber	$\$ = 89.217 X^\wedge 0.6442 \quad X = Chamber\,Volume\text{-}GAL$	1,060	423,000
6	Liquid Alum Feed	$\$ = 699.78 X + 88526 \quad X = Liquid\,Feed\,cap.\text{-}Gal/h$	2	1,000
7	Dry Alum Feed	$\$ = 212.32 x + 73225 \quad X = Dry\,Alum.\,Feed\text{-}Lb/h$	10	5,070
8	Polymer Feed (Cationic)	$\$ = 13,662 X + 20,861 \quad X = Polymer\,Feed\text{-}Lb/d$	1	220
9	Lime Feed	$\$ = 12.985 X^\wedge 0.5901 \quad X = Lime\,Feed\text{-}Lb/d$	10	10,000
10	Potassium Permanganate Feed	$\$ = 22,385 X^\wedge 0.0664 \quad X = Dry\,potassium\,perm\,Feed\text{-}Lb/d$	1	500
11	Sulfuric Acid Feed	$\$ = 32.606 X + 26395 \quad X = Sulfuric\,Acid\,(93\%)\,Feed\text{-}Gal/d$	11	5,300
12	Sodium Hydroxide feed	$\$ = 118.68 X + 38701 \quad X = Dry\,Sodium\,Feed\text{-}Gal/d$	9	10,000
13	Ferric Chloride Feed	$\$ = 20,990 X^\wedge 0.3190 \quad X = Dry\,Ferric\,Chlor.\,Feed\text{-}Lb/d$	13	6,600
14	Anhydrous Ammonia Feed (option)	$\$ = 7,959 X^\wedge 0.4235 \quad X = Ammonia\,Feed\text{-}Lb/d$	240	5,080
15	Aqua Ammonia Feed (Option)	$\$ = 3,014 X^\wedge 0.4219 \quad X = Ammonia\,Feed\text{-}Gal/d$	240	5,080
16	Powdered Activated Carbon	$\$ = 102,625 X^\wedge 0.2028 \quad X = Carbon\,Feed\text{-}Lb/h$	3	6,600
17	Rapid Mix G = 300	$\$ = 3.2559 X + 31023 \quad X = Basin\,Volume\text{-}GAL$	800	145,000
18	Rapid Mix G = 600	$\$ = 4.0668 X + 33040 \quad X = Basin\,Volume\text{-}GAL$	800	145,000
19	Rapid Mix G = 900	$\$ = 7.0814 X + 33269 \quad X = Basin\,Volume\text{-}GAL$	800	145,000
20	Flocculator G = 20 (10 minutes)	$\$ = 566045 X + 224745 \quad X = Basin\,Volume\text{-}MG$	0.02	7.00
21	Flocculator G = 50 (10 minutes)	$\$ = 673894 X + 217222 \quad X = Basin\,Volume\text{-}MG$	0.02	7.00
22	Flocculator G = 80 (10 minutes)	$\$ = 952,902 X + 177335 \quad X = Basin\,Volume\text{-}MG$	0.02	7.00
23	Circular Clarifier (10 ft walls)	$\$ = 2,989.8 X^\wedge 1.2346 \quad X = Diameter\text{-}LF$	30	200
24	Rectangular Clarifier	$\$ = 1.3572 X^\wedge 0.3182 \quad X = Basin\,Area\text{-}SF$	5,000	150,000
25	Gravity Filter Structure	$\$ = 15,338 X^\wedge 0.6499 \quad X = Filter\,Area\text{-}SF$	140	28,000
26	Filtration Media Stratified Sand (old design)	$\$ = 158 X + 11185 \quad X = Filter\,Media\,Area\text{-}SF$	140	28,000
27	Filtration Media Dual Media	$\$ = 38.319 X + 21377 \quad X = Filter\,Media\,Area\text{-}SF$	140	28,000
28	Filtration Media Mixed Media	$\$ = 62.844 X + 21838 \quad X = Filter\,Media\,Area\text{-}SF$	140	28,000
29	Filter Backwash Pumping	$\$ = 292.44 X + 92497 \quad X = Filter\,Surface\,Area\text{-}SF$	90	1,500
30	Surface Wash System Hydraulic	$\$ = 58.487 X + 69223 \quad X = Filter\,area\text{-}SF$	140	27,000

(Continued)

Table 5.2.1 (Continued)

No.	Process	Cost Equation	Min	Max
31	Air Scour Wash System	$ = 17,128 X^0.3864 X = Filter area-SF	140	27,000
32	Wash Water Surge Basin (Holding Tank)	$ = 50.157 X + 266176 X = Basin Capacity-GAL	9,250	476,000
33	Wash Water Storage Tank (Waste Holding)	$ = 5.6602 X^0.8473 X = Storage Volume-GAL	19,800	925,000
34	Clear Water Storage Below Ground	$ = 604450 X + 215121 X = Capacity-MG	0.01	8.00
35	Finished Water Pumping TDH-30.8 mts (100 ft)	$ = 57,887 X^0.7852 X = Pump Capacity MGD	1.45	300.00
36	Raw Water Pumping	$ = 12,169 X + 60716 X = Pump Capacity MGD	1.00	200.00
37	Gravity Sludge Thickener	$ = 4729.8 X + 37068 X = Thickener Diameter-FT	20	150
38	Sludge Dewatering lagoons	$ = 62792 X^0.7137 X = Storage Volume-MG	0.08	40.00
39	Sand Drying Beds	$ = 30.648 X^0.8751 X = Bed Area-SF	4,800	400,000
40	Filter Press	$ = 102794 X^0.4216 X = Filter Press Vol.-Gal/h	30	6,600
41	Belt Filter Press	$ = 146.29 X + 433972 X = Machine Capacity-Gal/h	800	53,000
42	Centrifuge Facility	$ = 328.03 X + 751295 X = Machine Capacity-Gal/h	1000	54,000
43	Administration, Laboratory, and Maintenance Building	$ = 63,568 X^0.553 X = Plant Capacity MGD	1	200

SUB TOTAL PROCESS COSTS
YARD PIPING 10%
SITEWORK LANDSCAPING 5%
SITE ELECTRICAL & CONTROLS 20%
TOTAL CONSTRUCTION COST
ENGINEERING, LEGAL & ADMINISTRATIVE COST 35%
TOTAL PROJECT COST

Figure 5.5.1a   One-Ton Chlorine Cylinders and Chlorine Feeder

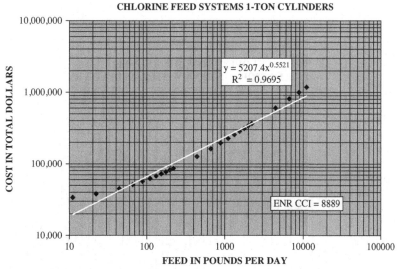

Figure 5.5.1b   One-Ton Chlorine Cylinders and Chlorine Feeder

When more than 200 lb/day of chlorine are required the design philosophy would require multiple1-ton cylinders resting horizontally on
load-cells and connected to a manifold and housed in an enclosure with
separate rooms for the cylinders and feed systems. Each room must be

mechanically ventilated at one complete air change per minute and in some areas must be serviced by an air scrubber to reduce the hazard of escaping chlorine gas into the local environment. Figure 5.5.1b is for the 1-ton system. The 1b designation is the reference number used in the general cost equation table shown earlier in this chapter. It and all the process cost curves will be in the appendix.

A 200 lb per day system shows a $100,000 construction cost, inclusive of the chlorinators, housing, ventilation, and process water supply. A 2,000 lb/day chlorination system would be about $340,000. These costs do not take into account the wide variety of architectural features necessary to enhance or minimize the visibility of the plant.

5.5.2 Chlorine Storage Tank with Rail Delivery or Feed from Rail Car

When design criteria permits, a larger storage and feed facility can be more cost-effective. A 20-ton storage tank can provide 8-days of chlorine at 5,000 lbs per day. (See Figure 5.5.2.)

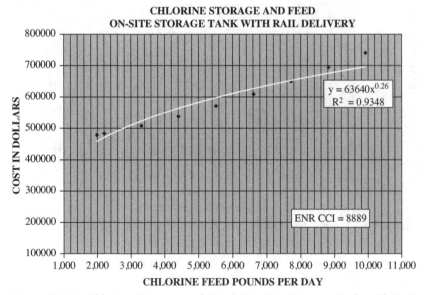

Figure 5.5.2 Chlorine Storage and Feed On-Site Storage Tank with Rail Delivery

5.5.3   Chlorine Direct Feed from Rail Car

And a direct feed from multiple rail cars on a siding can extend the time between deliveries thereby reducing the frequency and potential hazard of escaping chlorine gas during delivery. (See Figure 5.5.3.)

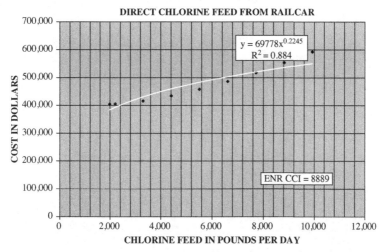

Figure 5.5.3   Direct Chlorine Feed from Railcar

5.5.4   Ozone Generation

Figure 5.5.4a   Ozone Generator

An ozone disinfection process is made up of a feed gas system, ozonator, contactor, and ozone destruct system. The ozone generation equipment for this cost curve includes: liquid oxygen (LOX) tanks, vaporizer and regulators, piping, valves flow meters, filters, and the ozone generators. Figure 5.5.4c represents the construction cost curve for the ozone generation process.

Figure 5.5.4b   Liquid Oxygen Storage Tank and Evaporators

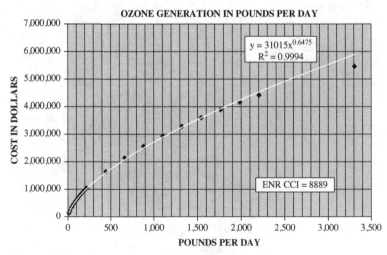

Figure 5.5.4c   Ozone Generation in Pound per Day

5.5.5   Ozone Contact Chamber

The ozone contactor is a separate process and is usually designed as a pair of contactors to allow for system maintenance and operations redundancy. The contactor is a cast-in-place concrete structure, including: piping, valves, diffusers, and an air handling system, which collects residual ozone and delivers it to the destruct unit. Figure 5.5.5 is the cost curve for the ozone contactors. If the design is for two contactors at full capacity, then the cost would be two times the cost for a single unit.

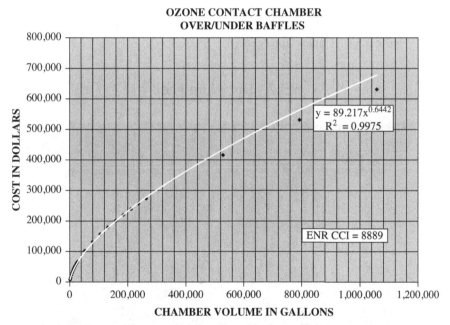

Figure 5.5.5   Ozone Contact Chamber Over/Under Baffles

5.5.6   Liquid Alum Feed

The chemical feed systems for liquid aluminum sulfate (Alum) consist of storage tanks, transfer pumps, metering pumps, piping and valves, and the facility enclosure. Figure 5.5.6 represents the cost of construction of these facilities over the range of 2 to 1,000 gal/hr.

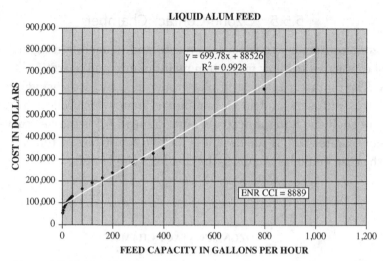

Figure 5.5.6 Liquid Alum Feed

5.5.7 Dry Alum Feed

Figure 5.5.7a Alum-Polymer Storage Tank

Figure 5.5.7b Dry Alum Feed includes: dry storage and feed, dissolving and mixing tank, metering and monitoring instruments, piping and valves, and a liquid metering and feed system for the liquid product.

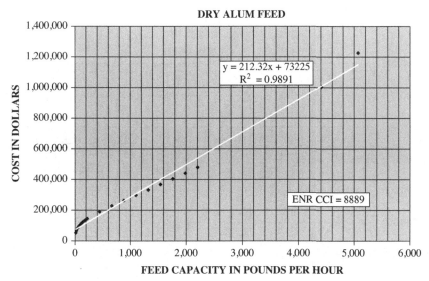

Figure 5.5.7b   Dry Alum Feed

5.5.8   Polymer Feed

Dry Polymer system design is more complex than the simple dry chemical feed system. It contains the same elements of the dry alum process above but also includes: a wetting unit, aging tank, transfer pumps, day tank, metering pumps, piping and valves, meters, and injectors. Figure 5.5.8 Polymer Feed shows the cost curve for the construction of this process.

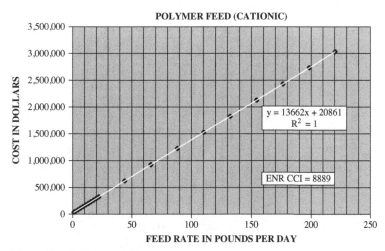

Figure 5.5.8   Polymer Feed

5.5.9   Lime Feed

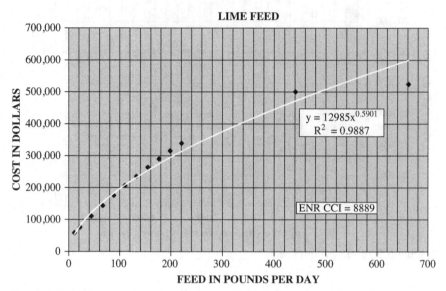

Figure 5.5.9   Lime Feed

The lime (quick lime) feed system includes a lime silo with bin activator and dust collector, a gravimetric dry chemical feeder, a slaker to prepare the lime slurry, and all other piping, valves, and meters. Figure 5.5.9 illustrates the construction cost of this facility over a range of 10 to 700 pounds per day.

5.5.10   Potassium Permanganate Feed (KMNO4)

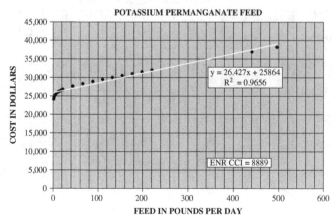

Figure 5.5.10   Potassium Permanganate Feed

Potassium Permanganate is a dry chemical and is stored and fed much like the dry alum and includes: dry storage and feed, dissolving and mixing tank, metering and monitoring instruments, piping and valves, and a hydraulic injector for the liquid product.

5.5.11   Sulfuric Acid Feed

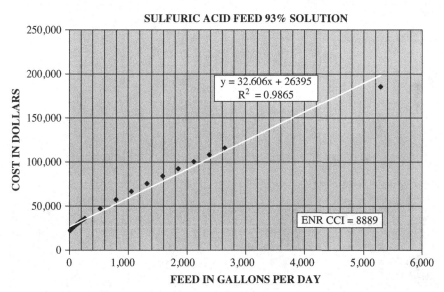

Figure 5.5.11   Sulfuric Acid Feed 93% Solution

Sulfuric acid is delivered in a liquid form at 93%. Its storage and feed is more hazardous to store and handle than some other requires separate secondary containment protection from other reactive chemicals. Figure 5.5.11 illustrates the construction cost of this process, including secondary containment.

5.5.12   Sodium Hydroxide Feed

Sodium hydroxide is usually delivered and stored in liquid form at 50%. This solution will freeze at 53 degrees Fahrenheit. Heat tracing

Figure 5.5.12a   Sodium Hydroxide Storage

and insulation for the storage tanks are recommended for cold climates. It requires secondary containment separate from other reactive chemicals like sulfuric acid. And its elements include: storage tanks, transfer pumps, metering pumps, piping and valves, and the facility enclosure. Figure 5.5.12b represents the construction cost of this process.

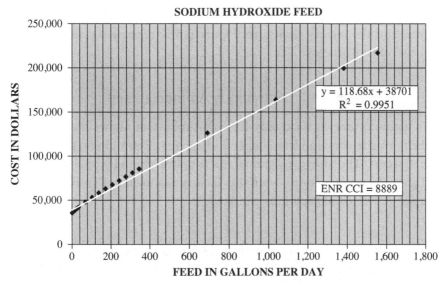

Figure 5.5.12b   Sodium Hydroxide Feed

5.5.13   Ferric Chloride Feed

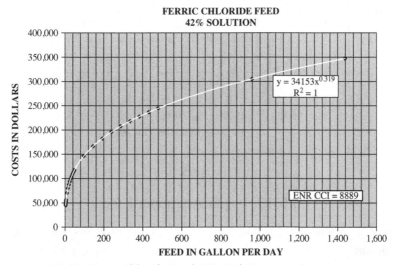

Figure 5.5.13   Ferric Chloride Feed 42% Solution

Ferric chloride is delivered in a liquid form at 42% solution. This solution will freeze below 20 degrees Fahrenheit. The range of application is from 2 to 1,450 gallons per day of liquid ferric chloride. Constructed facilities include: storage tanks, transfer pumps, metering pumps, piping and valves, and the facility enclosure. Figure 5.5.13 represents the construction cost of this process.

5.5.14 Anhydrous Ammonia Feed

This optional process used only for chloramination allows the ammonia and chlorine to combine for a longer-lasting disinfectant in large distribution systems. The anhydrous ammonia is delivered and stored as a liquefied gas. The construction cost of this facility is shown in Figure 5.5.14 for a range of 200 to 5,000 pounds per day.

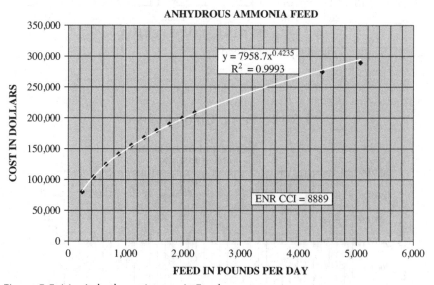

Figure 5.5.14   Anhydrous Ammonia Feed

5.5.15 Aqua Ammonia Feed

Aqua ammonia is delivered and stored as a liquid at 29% solution. Its use and performance is much the same as the anhydrous ammonia. It is stored and fed similarly to other liquid chemicals. Because it is unstable and strongly alkaline, it must have separate, secondary containment protection. Figure 5.5.15 illustrates the construction cost of the facility for a range from 50 to 750 gallons per day.

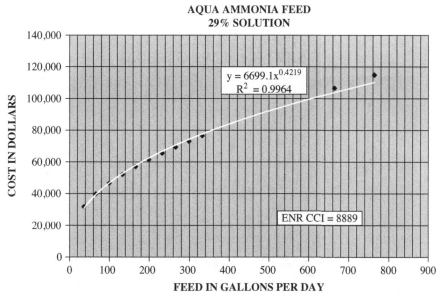

Figure 5.5.15   Aqua Ammonia Feed

5.5.16   Powdered Activated Carbon

Powdered activated carbon (PAC) conforms to the dry feed process model. It has a dry feeder with a bag-loading hopper, extension hopper, dust

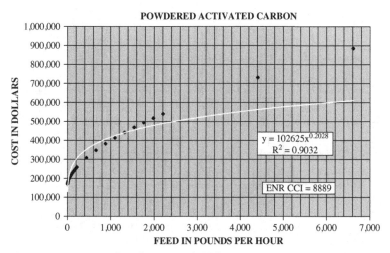

Figure 5.5.16   Powdered Activated Carbon

collector, and either a dissolving tank or a vortex mixer tank. Figure 5.5.16 shows the construction cost of the process. The curve shape is a polynomial of the form $y = ax^2 + bx + c$, where x is the amount of carbon expended in pounds per hour.

5.5.17   Rapid Mix G = 300

Figure 5.5.17a   Flocculation and Rectangular Clarifier Basins

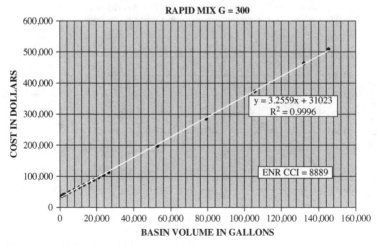

Figure 5.5.17b   Rapid Mix G = 300

Although the energy input for this process is considered too low for an effective process, it is included for comparison purposes. The second and third graph in the Rapid Mix series illustrates the construction costs for $G = 600$ and $G = 900$.

5.5.18   Rapid Mix G = 600

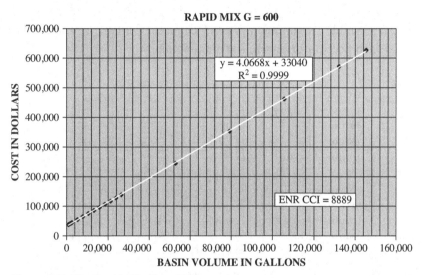

Figure 5.5.18   Rapid Mix G = 600

5.5.19   Rapid Mix G = 900

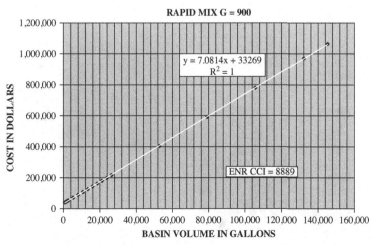

Figure 5.5.19   Rapid Mix G = 900

5.5.20   Flocculator G = 20

The flocculation basins consist of a linear set of three cast-in-place concrete basin with baffle walls separating them. Each basin has a single flocculator motor and gear box mounted on crossbeams above the basin with a vertical drive shaft ending in mixing blades. The detention time for each basins is 10 minutes. The following three cost curves (5.5.20, 21, and 22) are for different G forces applied by the flocculator motor to the process water. If variable frequency drive (VFD) motor controls are needed add $15,000 each flocculator purchased and installed.

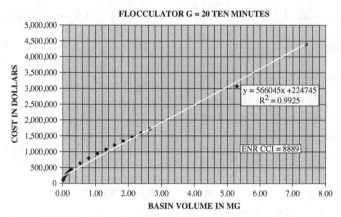

Figure 5.5.20   Flocculator G = 20

5.5.21   Flocculator G = 50

Figure 5.5.21a   Vertical Shaft Flocculator with Mixing Blades

Figure 5.5.21b   Horizontal Shaft Flocculator with Paddle-Type Blades

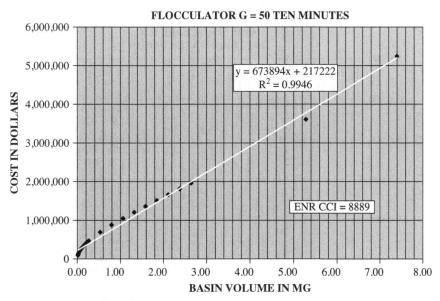

Figure 5.5.21c   Flocculator G = 50

5.5.22 Flocculator G = 80

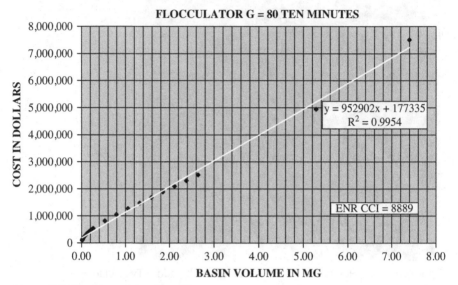

Figure 5.5.22   Flocculator G = 80

5.5.23 Circular Clarifier with 10-Ft Side Water Depth

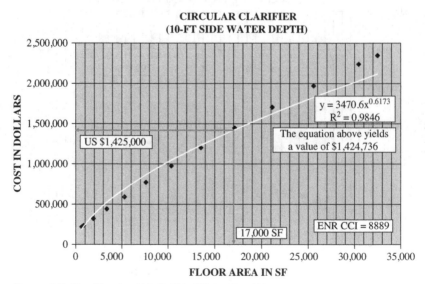

Figure 5.5.23   Circular (10-Ft Side Water Depth)

The circular clarifier is a cast-in-place concrete structure with a sloped bottom, and sludge rake mechanism, and center rake type sludge collector. Designed as an up-flow clarifier it has a central inlet line and an interior channel with a fixed weir around the perimeter wall to receive processed water. Figure 5.5.23 illustrates the construction cost of a partially buried circular clarifier.

5.5.24   Rectangular Clarifier

Figure 5.5.24a   Sedimentation Tank

Figure 5.5.24b   Center-Pivoted Rotating Rake Sludge Collector in Sedimentation Tank

The rectangular clarifier is a horizontal flow clarifier suited to larger municipal facilities. Rectangular clarifiers are usually designed in parallel modules to minimize the process footprint and to take advantage of the common wall and use a common inlet and outlet channel for process water. This clarifier is designed with a chain and flight sludge collection system with upper weir collection troughs for process water. Figure 5.5.24c illustrates the construction cost of this process.

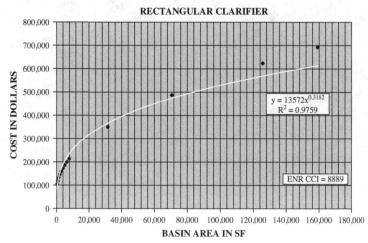

Figure 5.5.24c   Rectangular Clarifier

5.5.25   Gravity Filter Structure

In this manual the gravity filter structure is separate from the filter media in order to allow cost comparison between design issues and media types. The filter structure is a cast-in-place concrete structure with an

Figure 5.5.25a   Filter Pipe Gallery of the F. E. Weymouth Filtration Plant of MWD of Southern California

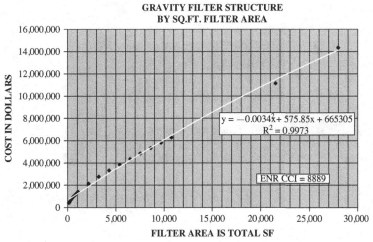

Figure 5.5.25b   Gravity Filter Structure by Sq. Ft. Filter Area

inlet channel, motor controlled butterfly valves, effluent gullet, and underdrain system. For cost-efficiency, pairs of filters are placed opposite each other with a common central gallery for piping and controls. Special consideration should be given to the number of filter cells to adequately accommodate the design and operation requirements and to plan for future expansion. Figure 5.5.25b illustrates the construction cost for this type of filter structure. The equation function for this cost curve, like the PAC is a polynomial of the form $y = ax^2 + bx + c$, where x is the square feet of filter area, making up a relatively small portion of the process footprint.

Figure 5.5.26a   Filter Cell, Granular Media Gravity Filter

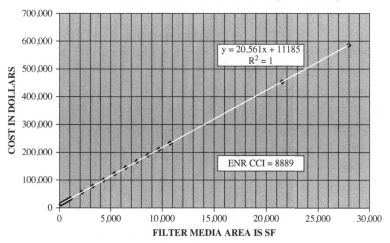

Figure 5.5.26b Filtration Media-Stratified Sand (Old Design)

5.5.26 Filter Media – Stratified Sand

The next three cost curves (see Figures 5.5.26b, 5.5.27, and 5.5.28) are for three basic types of filter media. The first, stratified sand, is rarely used but is shown for comparison purposes. The cost curves are linear since the cost includes the media material and the labor to place it.

5.5.27 Filter Media – Dual Media

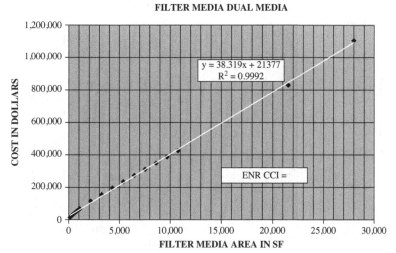

Figure 5.5.27 Filter Media Dual Media

5.5.28 Filter Multi-Media

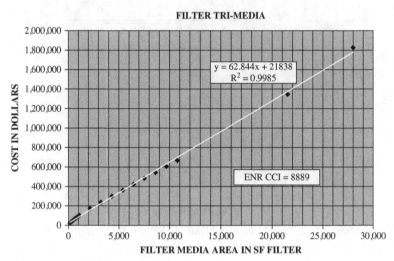

Figure 5.5.28   Filter Tri-Media

5.5.29   Filter Backwash Pumping

In the event that an elevated tank is not used filter backwash pumps are used. The filter media is backwashed using stored backwash water and high-flow, low-head pumps to lift the media and allow the unwanted

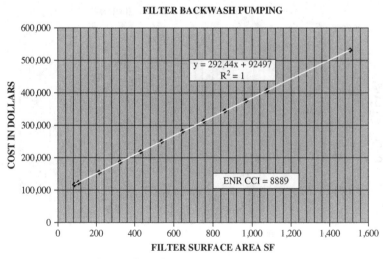

Figure 5.5.29   Filter Backwash Pumping

particles to overflow into the backwash troughs at the surface of the filter. The backwash pump station is usually a separate facility although adjacent to the wash water storage basins. Figure 5.5.29 represents the construction cost of this pump station including equipment, piping, and valves.

5.5.30   Surface Wash System

The surface wash system is made up of a pair of rotating arms (for each filter cell) that spray water just under the surface of the top 6 inches of filter media to loosen the upper layer of filtered sediment. Figure 5.5.30 illustrates the construction cost of the surface wash system.

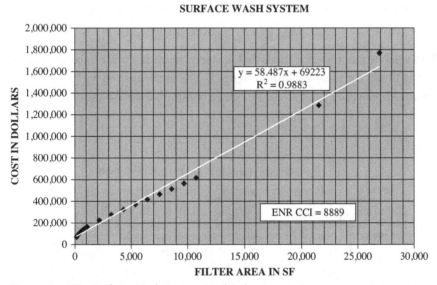

Figure 5.5.30   Surface Wash System Hydraulic

5.5.31   Air Scour Wash System

The air scour wash encompasses the entire volume of filter media. Air is entrained to assist in separating the media particles and allowing a more thorough cleaning. (See Figure 5.5.31.)

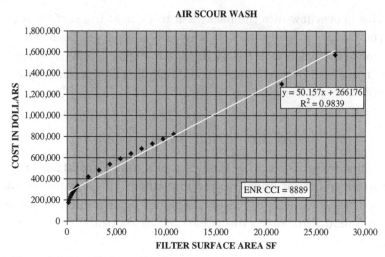

Figure 5.5.31 Air Scour Wash

5.5.32 Wash Water Surge Basin

The wash water used for the filters is stored in basins prior to being pumped to the filters for filter washing. Figure 5.5.32 is the cost curve for this surge basin. An elevated tank is used in place of filter backwash pumps.

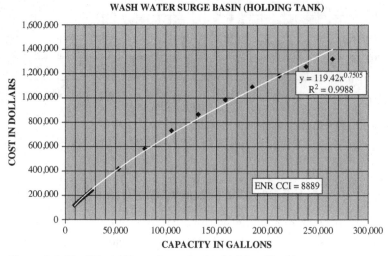

Figure 5.5.32 Wash Water Surge Basin (Holding Tank)

5.5.33   Filter Waste Wash Water Storage Tank

Figure 5.5.33a   Filter Waste Wash Water Storage Tank

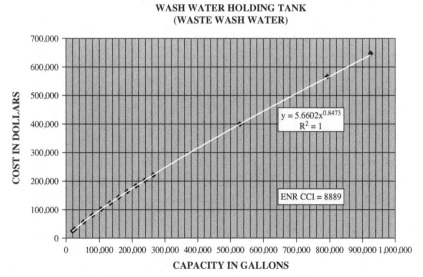

WASH WATER HOLDING TANK
(WASTE WASH WATER)

$y = 5.6602x^{0.8473}$
$R^2 = 1$

ENR CCI = 8889

COST IN DOLLARS

CAPACITY IN GALLONS

Figure 5.5.33b   Filter Waste Wash Water Storage Tank

The wash waste storage tank is usually designed to hold twice the volume of the surge tank so that it can hold multiple backwash cycles. This type of tank does is usually buried. The cost curve shown in Figure 5.5.33b shows the range over which the tank is set.

5.5.34   Clearwell Water Storage – Below Ground

This structure provides a buffer between the output of the treatment plant and the distribution system demand. It is also referred to as the clearwell. This is a large cast-in-place concrete structure covered by a concrete roof with interior supporting columns. Interior baffles are used to minimize short circuiting of the product water that could compromise the disinfection process. Figure 5.5.34 illustrates the construction cost of the clear water storage reservoir. The cost curve is based on easily excavated soil with no piles or groundwater problems. In the event that these issues arise the additional costs could be higher by 50% or more for this range of volume.

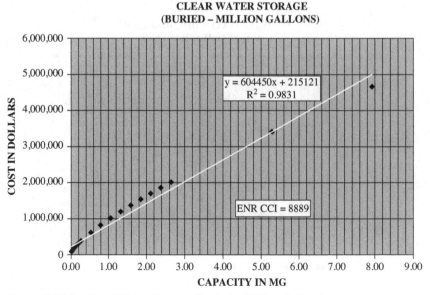

**CLEAR WATER STORAGE
(BURIED – MILLION GALLONS)**

$y = 604450x + 215121$
$R^2 = 0.9831$

ENR CCI = 8889

Figure 5.5.34   Clear Water Storage (Buried – Million Gallons)

5.5.35   Finish Water Pumping – TDH – 100 ft

Figure 5.5.35a   Finished Water Pumping Station (Centrifugal Pumps)

The finish water pumping station must be sufficiently sized to allow for operational redundancy and to provide the necessary downstream pressure. The typical design on the West Coast is for multiple vertical turbine pumps to draw water from the clearwell directly and pump the water into the distribution system on demand. (See Figure 5.5.35a.)

In the eastern United States a wet well/dry pit with horizontal centrifugal pump facility is often used. The cost of this type of facility can easily be 2 to 3 times higher than what is shown on the curve below. Other more

**FINISHED WATER PUMPING
(TDH 100 FT)**

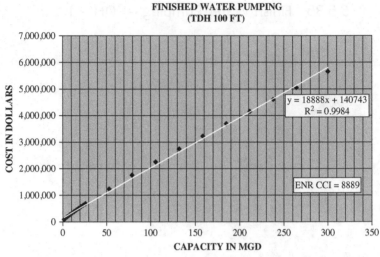

Figure 5.5.35b   Finished Water Pumping (TDH 100 Ft)

sophisticated pumping systems with multiple pressure zones should be priced using more precise design information.

5.5.36   Raw Water Pumping

Figure 5.5.36a   Six Vertical Pumps in Front of a Building

It is often necessary to lift the raw water higher than the source of supply so the treatment processes have sufficient gravity flow to overcome the head loss of the total plant and provide sufficient water to meet demand. This is usually a high-flow low-head pumping requirement to lift the raw water 20 to 30 ft. Figure 5.5.36b provides the cost curve for a simple pump station at the head of the plant.

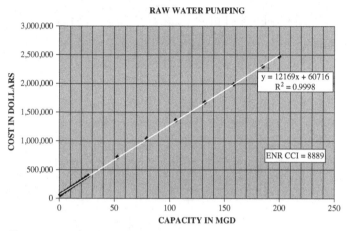

Figure 5.5.36b   Raw Water Pumping

5.5.37   Gravity Sludge Thickener

Figure 5.5.37a   Gravity Sludge Thickener

The gravity sludge thickener provides a sludge removal to separate the filter backwash and clarifier sludge for drying and ultimately disposal at a land fill. The process is much like the circular clarifier where the sludge is periodically pumped to sludge lagoons or drying beds. Figure 5.5.37b provides construction costs for a range of thickeners.

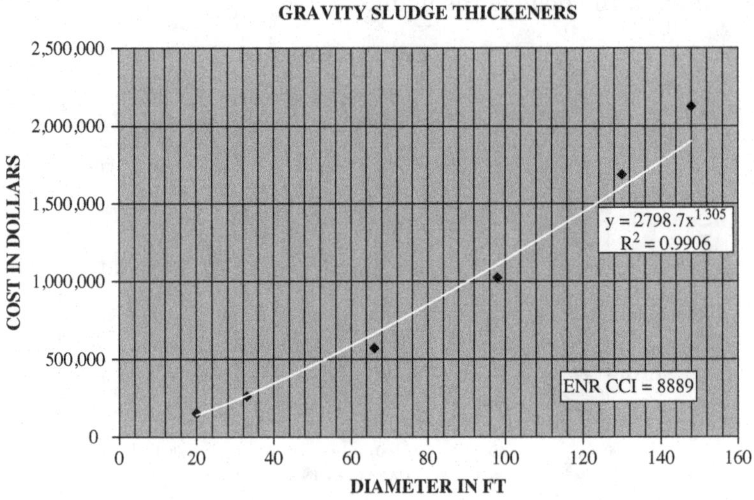

GRAVITY SLUDGE THICKENERS

$y = 2798.7x^{1.305}$
$R^2 = 0.9906$

ENR CCI = 8889

Figure 5.5.37b   Gravity Sludge Thickeners

5.5.38   Sludge Dewatering Lagoons

Figure 5.5.38a   View of Sludge Dewatering Lagoons

Sludge dewatering lagoons are constructed in soil by excavating and constructing berms to separate adjacent lagoons. The volume of cut and fill and the interconnecting sludge lines at the lagoons are the primary elements of the work. Figure 5.5.38c provides the cost curve for these lagoons.

Figure 5.5.38b   View of Sludge Dewatering Lagoon with Dried Sludge

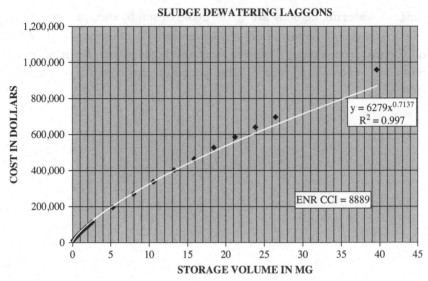

Figure 5.5.38c   Sludge Dewatering Lagoons

5.5.39   Sand Drying Beds

Figure 5.5.39a   View of Sand Drying Beds

Sludge drying is also accomplished by the use of sand drying beds. The sand allows the water to separate from the sludge and be collected by an underdrain system made of an impervious underlayment and perforated drainage piping to a collection system, where it is sent either back to the plant headworks, sewer or an adjacent leach field.

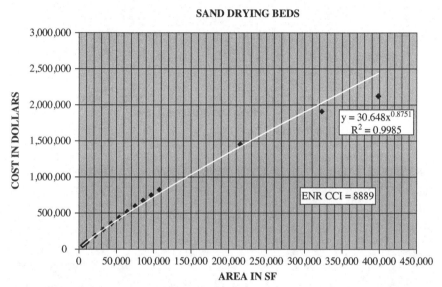

Figure 5.5.39b   Sand Drying Beds

5.5.40   Filter Press

Sludge can also be processed by the use of a filter press. When there is no land area to use either lagoons or sand drying beds to process the sludge, there are more expensive mechanical processes that can be used. Filter press pumps the sludge through a hollow multi-plate press separated by filter membranes to separate out the sludge from the liquid. The operation of the filter press can also be a significant cost compared with the drying lagoons pr sand beds. Figure 5.5.40 illustrates the construction cost of a typical facility.

5.5.41   Belt Filter Press

Another mechanical sludge-drying system is the belt filter press. This piece of equipment is less labor intensive than the plate press but is more

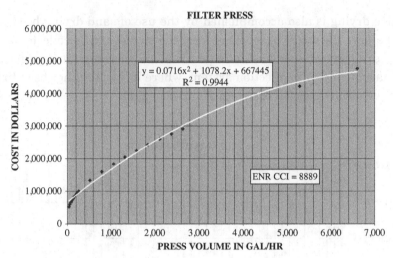

Figure 5.5.40   Filter Press

Figure 5.5.41a   View of Filter Belt Press

expensive to purchase and install. The belt press uses multiple perforated belts that mechanically compress the sludge, allowing the water
to run through the belts and to the decant tank, where it is sent to the
sewer, headworks, or leach field.

FILTER BELT PRESS

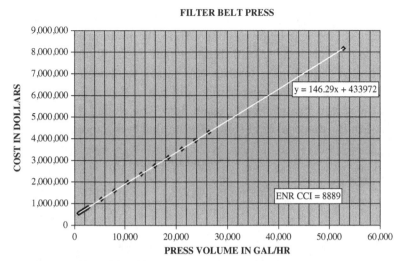

Figure 5.5.41b   Filter Belt Press

5.5.42   Centrifuge Facility

Figure 5.5.42a   Centrifuge Facility

Another mechanical sludge drying system is the centrifuge facility. This facility is usually multilevel with the centrifuges on the upper level and sludge disposal equipment below. The centrifuge facility is cleaner but more expensive than either the belt or filter press alternatives. The operation and maintenance costs are similar to the belt press with the equipment, being less labor intensive than the plate press but more expensive to purchase and install than the belt press. Sludge is pumped to the spinning centrifuges, forcing the water to flow from the sludge through the outside of the centrifuge and to the decant tank, where it is sent to the sewer, headworks, or leach field. The sludge is scavenged and conveyed to the disposal bin below to be hauled to a disposal site.

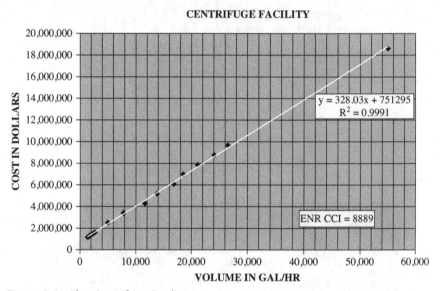

Figure 5.5.42b   Centrifuge Facility

5.5.43   Administration, Laboratory, and Maintenance Building

The treatment plant will typically have operations facilities where the plant business, maintenance, and testing can be performed. These are designed for separate purposes and have very different costs. Figure 5.5.43 compiles the cost of operations buildings into a single curve of average costs. And for this level of cost estimating this is good enough.

In using this curve, it is applied separately to each building. Administration will house the offices, reception area, restrooms, public spaces, and parking. The laboratory will have all the testing equipment, mechanical piping, hazardous and special containment facilities, separate wash rooms and showers, and so on. The maintenance area will have hydraulic hoists, monorail or traveling bridge cranes, a paint room, and tools and equipment to maintain the plant facilities. If all three buildings are

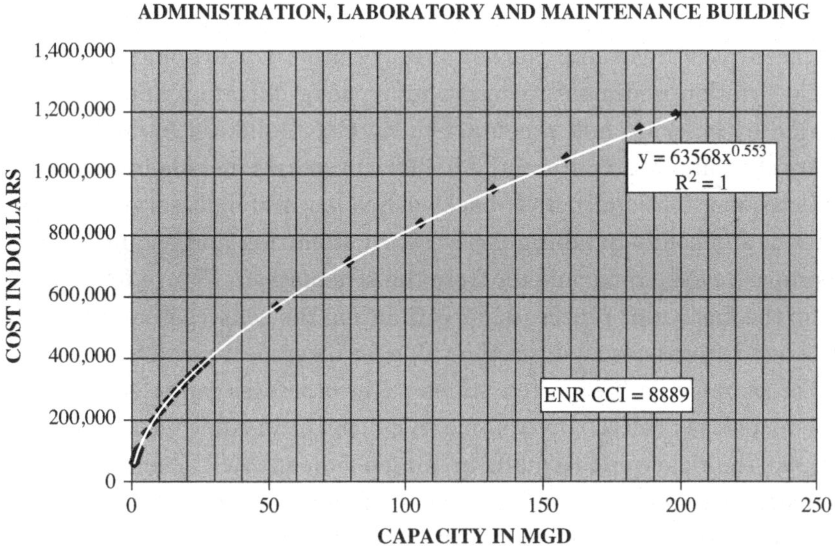

ADMINISTRATION, LABORATORY AND MAINTENANCE BUILDING

Figure 5.5.43   Administration, Laboratory, and Maintenance Building

necessary for a facility, the total cost will be three times the cost calculated by the equation or plotted on the curve at the intersection of the plant flow in million gallons per day, with the cost on the vertical, y-axis.

5.6 ESTIMATING CAPITAL COSTS

Using the design approach summarized in Chapter 2, it is important that care be taken even at the predesign level. The first four design commandments call for the following activities:

1. You shall make a careful analysis and evaluation of the quality of both raw and required finished waters.
2. You shall undertake a through evaluation of local conditions.
3. The treatment system developed shall be simple, reliable, effective, and consist of proven treatment processes.
4. The system considered shall be reasonably conservative and cost-effective.

This will require some engineering and process calculations before providing cost estimates to the owner and other stakeholders. When compiled, the separate process costs can be summed and the cost of the entire treatment plant estimated.

As illustrated in the schematic below a conventional process train for surface water treatment consist of coagulation with rapid mix followed by flocculation, sedimentation, granular media filtration and final disinfection by chlorine and a contact tank (**C*t**), followed by at least four hours of treated water storage. Ancillary processes include intake screen, grit chamber (optional), filter backwash, a low and high-service pumping station, and solids-handling facilities. By itemizing and establishing the appropriate design parameter from the schematic in Figure 5.6.1, we develop the basic unit processes for estimating the construction cost. Along with each process, we will set the value for its respective parameter.

For example, in the United States water processes can be set at million gallons per day (MGD), by volume in gallons or in cubic feet. Water storage like the clearwell is usually in million gallons (MG). Chemical storage and feed is usually in pounds per day or pounds per hour with the storage volume sized for 20 days at average plant flow.

However, certain chemicals such as liquid alum, ferric chloride, caustic soda, cationic polymers, hydrofluosilic acid, and zinc orthophosphate are commercially available with in liquid form at specific strengths. In these cases, actual storage and feed rates are generally expressed as gallons, gallons per hour, or gallons per day.

The plant in this example has two process pumping facilities to transfer collected waste and solids to be further processed. There is a complete

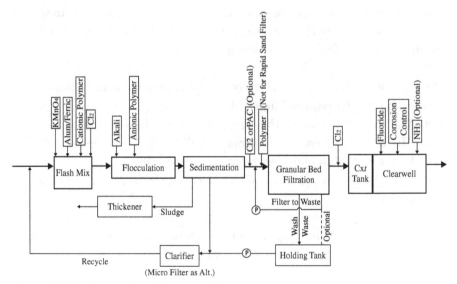

Figure 5.6.1   Conventional Treatment Process

solids-handling facility, including a solids holding tank, pumping station, clarifier, and sludge thickener.

5.7 ESTIMATING CAPITAL COSTS OF A CONVENTIONAL WATER TREATMENT PLANT

Table 5.7.1 is an itemization of the conventional treatment processes for an average daily flow rate of 100 MGD. Parametric values are set on a per module basis and a cost per module is calculated. This is typical of most design criteria where operation and maintenance requires that a portion of each process lie idle and under maintenance or going through a cleaning cycle such as filter backwash. The modular cost is multiplied by the number of process modules, and a total process cost is extended to the Total Cost column. These process costs are summed and percentage costs for yard piping, other sitework, and electrical and instrumentation work are calculated and summed to a Total Construction Cost. Engineering, Legal, and Administration costs are calculated as a percentage (35% of construction) and added for a Total Capital Cost of $158 million in 2007 dollars.

5.7.1   Two-Stage Filtration Plant

Figure 5.7.1 is a composite of the construction cost and nonconstruction costs for the *two-stage filtration* plant. The range of the treatment plant cost curve is from 10 MGD to 100 MGD and is made up of a selection of the individual process costs for the estimated process parameter. The process cost tables for 10 MGD and 100 MGD are in the Appendix.

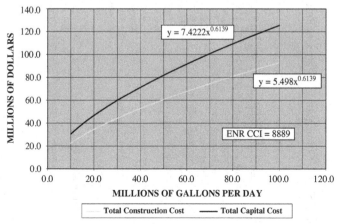

Figure 5.7.1   Two-Stage Filtration

Table 5.7.1   Conventional Treatment Processes for an Average Daily Flow of 100 MGD

No.	Process	Cost Equation	Applicable Range		Quantity Per Unit Process	Number of Process Units	Process Cost Per Unit	Total Process Cost
			Minimum	Maximum				
1a	Chlorine storage and feed 150# cylinder storage	$ = 1181.9X^0.06711·X = Chlorine feed cap.-Lb/d	10	200			$0	$0
1b	Chlorine storage and feed 1-ton cylinder storage	$ = 5207.41X^0.6621·X = Chlorine feed cap.-Lb/d	200	10,000	2000	2	$985,458	$1,970,917
2	On-site storage tank with rail delivery	$ = 6340X^0.2600·X = Chlorine feed cap.-Lb/d	1,980	10,000			$0	$0
3	Direct feed from rail car	$ = 69778X^0.2245·X = Chlorine feed cap.-Lb/d	1,980	10,000			$0	$0
4	Ozone Generation	$ = 31,015X^0.6475·X = Ozone feed cap.-Lb/d	10	3,500			$0	$0
5	Ozone Contact Chamber	$ = 89,217X^0.6442·X = Chamber Volume-GAL	1,060	423,000			$0	$0
6	Liquid Alum Feed - 50% Solution	$ = 699.78X + 88526·X = Liquid Feed cap.-Gal/h	2	1,000	35	2	$139,493	$278,986
7	Dry Alum Feed	$ = 212.32x + 73225·X = Dry Alum. Feed-Lb/d	10	5,070	35	2	$99,550	$199,100
8	Polymer Feed	$ = 13662X + 20861·X = Polymer Feed-Lb/d	1	220	20	2	$362,998	$725,996
9	Lime Feed	$ = 12985X^0.5901·X = Lime Feed-Lb/d	10	10,000	400	2	$549,945	$1,099,890
10	Potassium Permanganate Feed	$ = 26.427X + 25864·X = Dry KMNO4 Feed-Lb/d	1	500	40	1	$33,228	$33,228
11	Sulfuric Acid Feed - 93% Solution	$ = 32.606X + 26395·X = Sulfuric acid (93%) Feed-Gal/d	11	5,300			$0	$0

No.	Description	Equation						
12	Sodium Hydroxide Feed - 50% Solution	$\$ = 118.68\,X + 38701 \cdot X =$ Liquid Sodium Feed-gal/d	1	1,600	400	2	\$144,716	\$289,432
13	Ferric Chloride Feed - 42% Solution	$\$ = 20990\,X^{0.3190} \cdot X =$ Dry Ferric Sulf. Feed-Lb/d	13	6,600			\$0	\$0
14	Anhydrous Ammonia Feed - 29% Solution	$\$ = 7959\,X^{0.4235} \cdot X =$ Ammonia Feed-Lb/d	240	5,080			\$0	\$0
15	Aqua Ammonia Feed	$\$ = 6699\,X^{0.4219} \cdot X =$ Ammonia Feed-Gal/d	240	5,080			\$0	\$0
16	Powdered Activated Carbon	$\$ = 0.0142X^2 + 195.03X + 194823 \cdot X =$ Carbon Feed-Lb/d	3	6,600	40	1	\$250,177	\$250,177
17	Rapid Mix G = 300	$\$ = 3.2559\,X + 31023 \cdot X =$ Basin Volume-GAL	800	145,000			\$0	\$0
18	Rapid Mix G = 600	$\$ = 4.0668\,X + 33040 \cdot X =$ Basin Volume-GAL	800	145,000	1750	1	\$49,587	\$49,587
19	Rapid Mix G = 900	$\$ = 7.0814\,X + 33269 \cdot X =$ Basin Volume-GAL	800	145,000	1750	1	\$56,383	\$56,383
20	Flocculator G = 20	$\$ = 566045\,X + 224745 \cdot X =$ Basin Volume-MG	0.015	7			\$0	\$0
21	Flocculator G = 50	$\$ = 673894\,X + 217222 \cdot X =$ Basin Volume-MG	0.015	7	0.035	2	\$297,357	\$594,713
22	Flocculator G = 80	$\$ = 952902\,X + 177335 \cdot X =$ Basin Volume-MG	0.015	7	0.035	2	\$260,163	\$520,325
23	Circular Clarifier	$\$ = 3470.6\,X^{0.6173} \cdot X =$ Basin Area-SF	650	32,300			\$0	\$0
24	Rectangular Clarifier	$\$ = 13572\,X^{0.3182} \cdot X =$ Basin Area-SF	5,000	150,000			\$0	\$0
25	Gravity Filter Structure	$\$ = -0.0034X^2 + 575.85X + 665305 \cdot X =$ Filter Area-SF	140	28,000	438	4	\$1,131,829	\$4,527,318
26	Filtration Media - Stratified Sand	$\$ = 20.561\,X + 11185 \cdot X =$ Filter Media Area-SF	140	28,000			\$0	\$0
27	Filtration Media - Dual Media	$\$ = 38.319\,X + 21377 \cdot X =$ Filter Media Area-SF	140	28,000	400	4	\$45,323	\$181,294

(Continued)

77

Table 5.7.1   (Continued)

No.	Process	Cost Equation	Applicable Range Minimum	Applicable Range Maximum	Quantity Per Unit Process	Number of Process Units	Process Cost per Unit	Total Process Cost
28	Filtration Media - Tri - Media	$ = 62.844X + 21838 - X =$ *Filter Media Area-SF*	140	28,000			$0	$0
29	Filter Backwash Pumping	$ = 292.44X + 92497 - X =$ *Filter Surface area-SF*	90	1,500	200	2	$186,458	$372,916
30	Surface Wash System	$ = 58.487X + 69223 - X =$ *Filter area-SF*	140	27,000	200	2	$99,941	$199,881
31	Air Scour Wash	$ = 50.157X + 266176 - X =$ *Filter area-SF*	140	27,000	200	2	$463,853	$927,706
32	Wash Water Surge Basin - (Holding Tank)	$ = 119.42X^{0.7505} - X =$ *Basin Capacity-GAL*	9,250	476,000	90000	1	$770,643	$770,643
33	Wash Water Storage Tank - (Waster Wash Water)	$ = 5.6602X^{0.8473} - X =$ *Storage Volume-GAL*	19,800	925,000	200000	1	$216,770	$216,770
34	Clear Water Storage - Below Ground	$ = 604450X + 215121 - X =$ *Capacity-MG*	0.011	8	1.700	1	$1,534,501	$1,534,501
35	Finished Water Pumping TDH = 100ft	$ = 18888x + 140743 - X =$ *Pump Capacity MGD*	1.45	300	4.66	3	$282,487	$847,462
36	Raw Water Pumping	$ = 13889x + 103488 - X =$ *Pump Capacity MGD*	1	200	4	3	$135,079	$405,238
37	Gravity Sludge Thickener	$ = 2798.7X^{1.305} - X =$ *Thickener Diameter-FT*	20	150	10	1	$94,864	$94,864
38	Sludge Dewatering lagoons	$ = 62792X^{0.7137} - X =$ *Storage Volume-MG*	0.08	40	0.02	3	$4,173	$12,519
39	Sand Drying Beds	$ = 30.648X^{0.8751} - X =$ *Bed Area-SF*	4,800	400,000	3333	6	$45,801	$274,805

No.	Item	Equation					Unit Cost	Total Cost
40	Filter Press	$\$ = -0.0716X^2 + 1078.2X + 667445 - X =$ / *Filter Press Vol.-GAL*	30		6,600		$0	$0
41	Belt Filter Press	$\$ = 146.29 X + 433972 - X =$ / *Machine Capacity-GAL/h*	800		53,000		$0	$0
42	Administration, Laboratory, and Maintenance Building	$\$ = 63568 X^{0.553} - X =$ / *Plant Capacity MGD*	1	200	10	1	$280,442	$280,442

SUB TOTAL PROCESS COSTS	$16,715,093
YARD PIPING 10%	$1,671,509
SITEWORK LANDSCAPING 5%	$835,755
SITE ELECTRICAL & CONTROLS 20%	$3,343,019
TOTAL CONSTRUCTION COST	$22,565,375
ENGINEERING, LEGAL & ADMINISTRATIVE COST 35%	$7,897,881
TOTAL PROJECT COST	$30,463,256

5.7.2   Direct Filtration Plant

Figure 5.7.2 is a composite of the construction cost and nonconstruction costs for the *direct filtration* plant. The range of the treatment plant cost curve is from 10 MGD to 100 MGD and is made up of a selection of the individual process costs for the estimated process parameter. The process cost tables for 10 MGD and 100 MGD are in the Appendix.

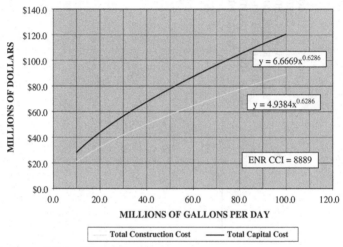

$y = 6.6669x^{0.6286}$

$y = 4.9384x^{0.6286}$

ENR CCI = 8889

MILLIONS OF GALLONS PER DAY

Total Construction Cost ——— Total Capital Cost

Figure 5.7.2   Direct Filtration

5.7.3   Conventional Filtration Plant

Figure 5.7.3a   New Mohawk Water Treatment Plant., Tulsa, Oklahoma

Figure 5.7.3b is a composite of the construction cost and nonconstruction costs for the *conventional filtration* plant. The range of the treatment plant cost curve is from 10 MGD to 100 MGD and is made up of a selection of the individual process costs for the estimated process parameter. The process cost tables for 10 MGD and 100 MGD are in the Appendix.

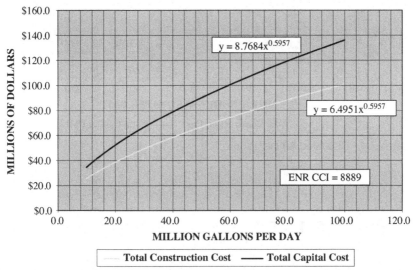

Figure 5.7.3b   Conventional Filtration

5.7.4   Dissolved Air Flotation Filtration Plant

Figure 5.7.4 is a composite of the construction cost and nonconstruction costs for the *dissolved air flotation filtration* plant. The range of the treatment plant cost curve is from 10 MGD to 100 MGD and is made up of a selection of the individual process costs for the estimated process parameter. The process cost tables for 10 MGD and 100 MGD are in the Appendix.

5.7.5   Lime and Soda Ash Filtration Plant

Figure 5.7.5 is a composite of the construction cost and nonconstruction costs for the *lime and soda ash filtration* plant. The range of the treatment plant cost curve is from 10 MGD to 100 MGD and is made up of a selection of the individual process costs for the estimated process parameter. The process cost tables for 10 MGD and 100 MGD are in the Appendix.

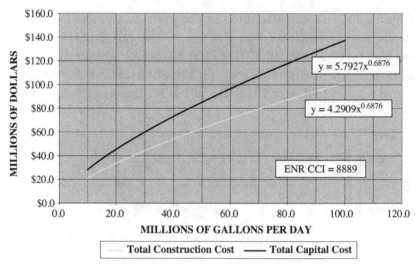

Figure 5.7.4   Dissolved Air Filtration

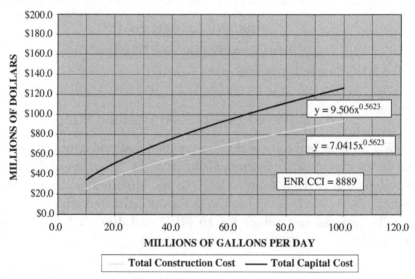

Figure 5.7.5   Lime and Soda Ash Filtration

5.7.6   Iron and Manganese Filtration Plant

Figure 5.7.6 is a composite of the construction cost and nonconstruction costs for the *iron manganese removal filtration* plant. The range of the treatment plant cost curve is from 10 MGD to 100 MGD and is made up of a selection of the individual process costs for the estimated process parameter. The process cost tables for 10 MGD and 100 MGD are in the Appendix.

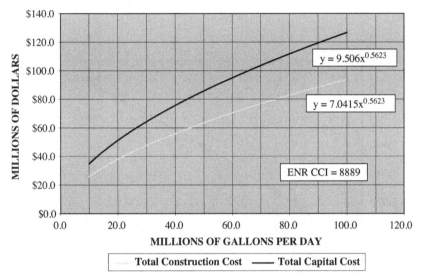

Figure 5.7.6   Iron Manganese Removal

5.7.7   Micro Membrane Filtration Plant

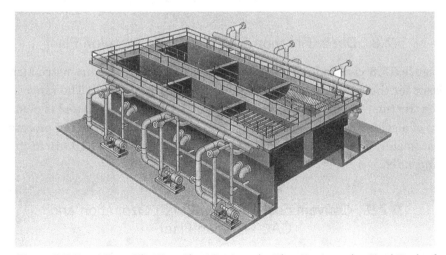

Figure 5.7.7a   Micro-Filtration Plant Designed with a Rectangular Steel Tank of Six Compartments

Figure 5.7.7b is a composite of the construction cost and nonconstruction costs for the *micro membrane filtration* plant. The range of the treatment plant cost curve is from 10 MGD to 100 MGD and is made up of a selection of the individual process costs for the estimated process parameter. The process cost tables for 10 MGD and 100 MGD are in the Appendix.

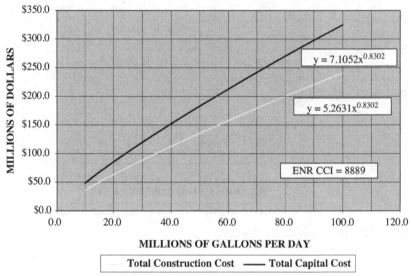

Figure 5.7.7b   Micro Membrane Filtration

5.7.8   Direct Filtration with Pre-ozone Filtration Plant

Figure 5.7.8 is a composite of the construction cost and nonconstruction costs for the *direct filtration with pre-ozone filtration* plant. The range of the treatment plant cost curve is from 10 MGD to 100 MGD and is made up of a selection of the individual process costs for the estimated process parameter. The process cost tables for 10 MGD and 100 MGD are in the Appendix.

5.7.9   Conventional Treatment with Ozonation and GAC Filtration Plant

Figure 5.7.9 is a composite of the construction cost and nonconstruction costs for the *conventional treatment with ozonation and GAC filters*. The

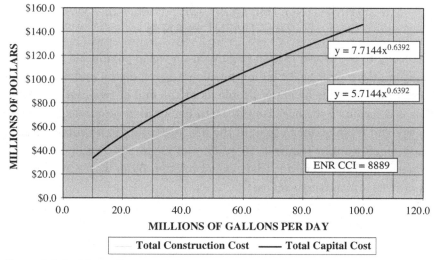

Figure 5.7.8   Direct Filtration with Pre-Ozone Construction Cost

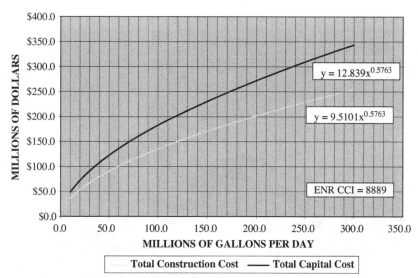

Figure 5.7.9   Conventional Treatment with Ozonation and GAC Filters Construction Cost

range of the treatment plant cost curve is from 10 MGD to 100 MGD and is made up of a selection of the individual process costs for the estimated process parameter. The process cost tables for 10 MGD and 100 MGD are in the Appendix.

5.8 ESTIMATING THE COST OF ADVANCED WATER TREATMENT PLANTS

According to the World Health Organization world population growth estimation, there will be 12 billion people on earth at the end of this century, which is twice as many people as at the end of twentieth century. This is a serious issue because the earth has limited resources, including energy and fresh water to sustain the life. According to a United Nations study, each human requires 15 gallons of clean safe water to sustain proper daily life. However, there is not enough fresh water available to support 12 billion people on earth.

Global warming trend makes situation worse. And the disproportionate transfer of water within coastal regions of developed countries could leave the interiors susceptible to drought and water scarcity. For example, the Southern California and the Greater New York metropolitan areas are currently supplying an average of 140

Figure 5.8a   Ion Exchange Demineralization Unit System

gallons per capita per day (gpcd), which is almost 10 times higher than 15 gpcd.

Therefore, advanced water treatment technologies are vitally important to world health. Many large cities in the world are built near the seashore giving the cities access to abundant seawater. Using a combination of advanced treatment such as; ultra-filtration (UF) and reverse osmosis (RO) as the primary water treatment processes can easily provide the required 15 gpcd of potable drinking water. Many African countries have serious water shortage problems, making prepackaged advanced water treatment plants like RO and UF highly cost-effective compared to disease and economic collapse from severe shortages of safe drinking water. Developed countries can easily supply the technology and equipment in this era of globalization, resulting in a mutually beneficial economic and social exchange with developing countries.

There are four seawater desalination treatment strategies currently in use with widely ranging construction and O&M costs. These treatment strategies are; reverse osmosis (RO), multi-stage flash (MSF), mechanical vapor compression (MVC), and multiple-effect distillation (MED). Figure 5.8b below shows the relative comparison of construction costs for these plants.

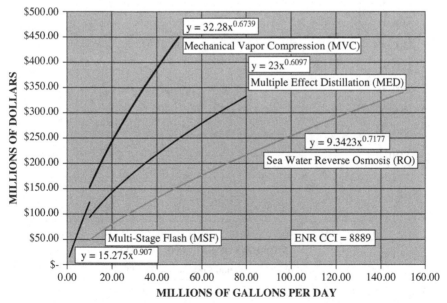

Figure 5.8b   Total Project Construction Cost Comparison

5.8.1   Reverse Osmosis (RO) Treatment Plant

Seawater desalination is currently provided by four alternative treatment processes. The most cost-effective of these is reverse osmosis. In general, pretreatment is accomplished by the use of a two-step process. During the first stage of the process a coagulant is added just prior to the clarifiers and allowed to settle before filtration. Conventional water and air backwash systems are included to maintain cleanliness of the filters. System uses sulfuric acid and a scale inhibitor in order to control scaling of the membrane surfaces.

Following pretreatment, the pressure is boosted to about 1,000 psig, water is fed to the membranes and product water is produced. The concentrate stream is then fed to an energy recovery device. This lowers total process energy use by about 30% - 40%. Recovery ratios are between 40% and 50%. Figure 5.8.1 below illustrates the construction cost of a treatment plant over a range of 10 MGD to 150 MGD of product water.

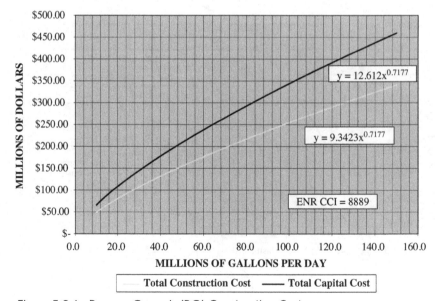

Figure 5.8.1   Reverse Osmosis (RO) Construction Cost

5.8.2   Multiple-Effect Distillation (MED) Treatment Plant

Multiple-effect distillation is the second most cost-effective treatment for seawater over the same range of 10 MGD to 50 MGD. The effect stream in the first effect is used as the heat source; it evaporates a small portion of the seawater entering the unit. The vapor produced is sent to the second effect, where it becomes the heat source for further evaporation. This procedure continues in the following effects until it reaches the last effect, where the final vapor is condensed in the main (or final) condenser. Figure 5.8.2 shows the construction cost for the treatment plant.

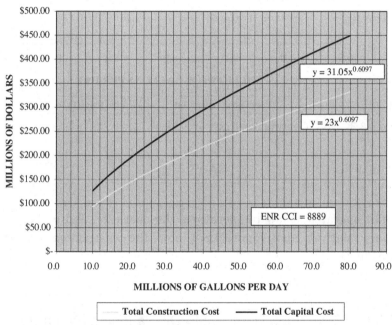

Figure 5.8.2   MED Construction Cost

5.8.3   Mechanical Vapor Compression (MVC) Treatment Plant

Mechanical vapor compression is an alternative treatment for seawater at flows below 10 MGD. In this treatment process, vapor produced in the evaporator is sent to a compressor. The compression raises the pressure and temperature enough so the vapor acts as the heat source for further evaporation. This process can achieve a recovery ration of 50%. Although

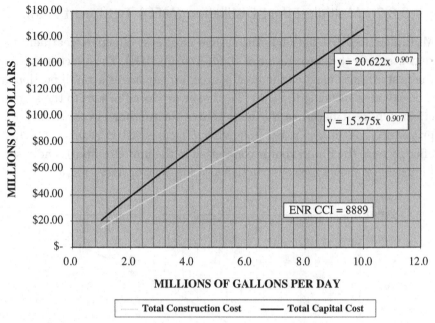

Figure 5.8.3   MVC Construction Cost

the unit operates at low temperature, the operation and maintenance costs are relatively high. Figure 5.8.3 shows the construction cost for the treatment plant.

5.8.4   Multi-Stage Flash (MSF) Distillation Treatment Plant

Multi-stage flash distillation is the highest-cost treatment for seawater over the same range of 10 MGD to 50 MGD. This treatment process uses high-temperature additives and a complex pretreatment process and re-circulation. The recovery section is composed of a number of flash chambers and heat exchangers in multiple stages. The recirculation stream condenses the vapors made in the flashing chamber. The recycle stream obtains its final temperature rise in the brine heater (which controls the final temperature of the process). The stream then reenters the first flash chamber where flashing begins again. Figure 5.8.4 shows the construction cost for the treatment plant.

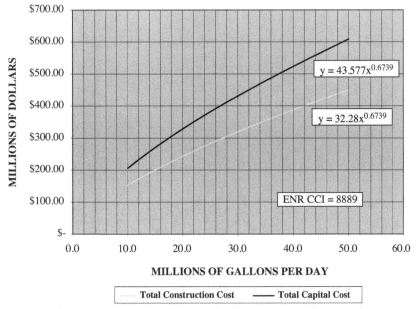

Figure 5.8.4   MSF Construction Cost

5.8.5   Ultra-filtration and Nano-Filtration

Ultra-filtration (UF) and *nano-filtration* (NF) have become reliable processes that fill the gap between low-pressure microfiltration and high-pressure reverse osmosis. The membranes and driving pressures have made these processes cost effective alternatives allowing a smaller, albeit more expensive RO process to provide the removal of ionized salts and other colloidal particles to be removed from source water at a lower operation and maintenance cost.

Ultra-filtration uses membranes significantly smaller (less than 0.1 mm) than those provided by the microfilters, while removing colloids, bacteria, viruses, and high-molecular-weight organic compounds (*Integrated Design and Operation of Water Treatment Facilities*, Kawamura, Wiley & Sons, 2000). Even at a pressure of 10 to 40 psig, these membranes are more susceptible to clogging and must be frequently backwashed.

Nano-filtration membranes are smaller yet (between 0.001 to 0.002 mm) with a pressure of 75 to 150 psig. Somewhere between the ultra-filtration size and the RO of (<1 nm) whose pressure is greater than 200 psig. The current preferred design is a train of UF/RO or NF/RO filters with other process elements providing the requisite water treatment and solids

Figure 5.8.5a   Membrane Filters

Figure 5.8.5b   Membrane Filters Fiber Details

handling required by regulators. The capital cost of these facilities varies significantly with the cost of the NF about 1.5 times that of RO. Costs for UF are usually in the upper third of the difference between Conventional Treatment and RO. Figure 5.8.5c below gives a range of probable costs for the combined facility.

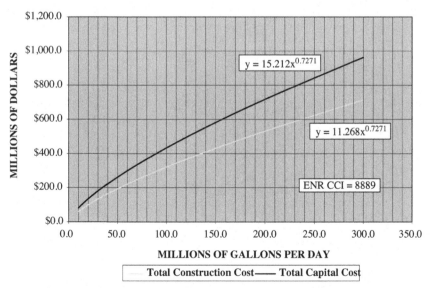

Figure 5.8.5c   UF NA Construction Cost

Chapter 6

Operation and Maintenance Cost Impacts

6.1 ANNUAL OPERATION AND MAINTENANCE COSTS

When developing the preliminary design report, it is often the case that a set of alternative water treatment plant types is evaluated. Along with the estimated construction and total capital cost, it is necessary to estimate the **annual** Operations & Maintenance (O&M) cost. The cost curves below were developed from data published by a number of public agencies and compiled into a database. The data was brought forward by the use of an index previously referred to as the ENR-Construction Cost Index. For the purpose of this manual we have set the ENR-CCI = 8889. All costs are in US dollars and all indexes were brought current to June 2007.

Operation and maintenance costs are made up of both fixed costs and variable costs. The fixed costs are typically labor, supervision, and administration. Variable costs are associated with chemicals, power, maintenance repairs and replacement of plant and equipment, and other supplies and services that are necessary to operate the process plant and supporting facilities. There are many additional factors that can significantly affect these costs, including the policies of the owner, climate and weather, from both a local plant site and the raw water source of surface waters. In addition, the sophistication of the plant instrumentation and controls can have a great effect on these costs.

O&M cost curves were developed for each of the common types of treatment plants identified in Chapter 2 and the advanced water treatment plants identified in Chapter 5.

Actual O&M costs can vary significantly between two plants with the same flow, treatment processes, and instrumentation even within the same owner/agency. And these cost curves should only be taken as a preliminary estimate when comparing a group of alternatives. When evaluating a single alternative, more care should be given to analyzing the owner/agency's policies and the system already in place.

6.2 O&M COST CURVES

O&M cost curves for the common treatment plants are shown below in the same order as in Chapter 2. These common plant O&M curves range in design flow between 10 MGD and 100 MGD.

6.2.1a   Two-Stage Filtration

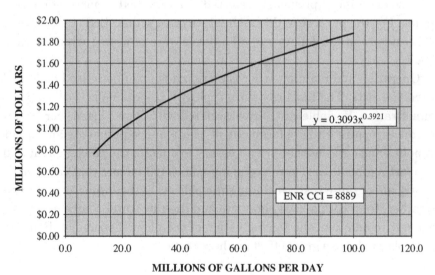

$$y = 0.3093x^{0.3921}$$

ENR CCI = 8889

Figure 6.2.1a   O&M Costs for Two-Stage Filtration

6.2.1b   Direct Filtration

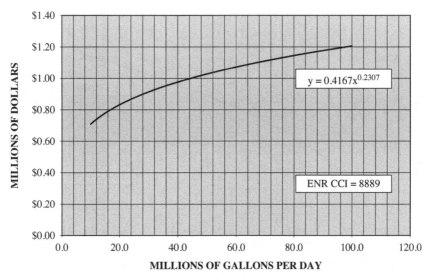

Figure 6.2.1b   O&M Costs for Direct Filtration

6.2.1c   Conventional Treatment

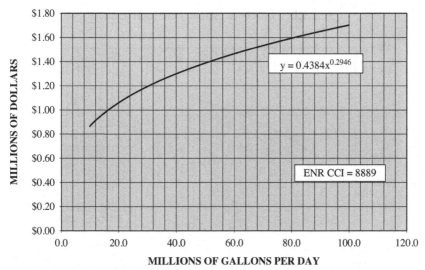

Figure 6.2.1c   O&M Costs for Conventional Treatment

6.2.2a   Dissolved Air Flotation

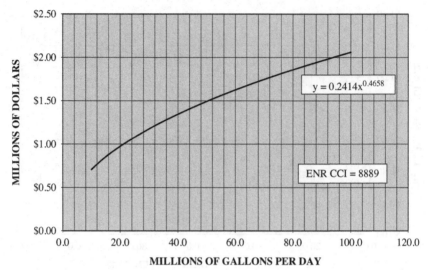

Figure 6.2.2a   O&M Costs for Dissolved Air Flotation

6.2.2b   Lime & Soda Ash Softening

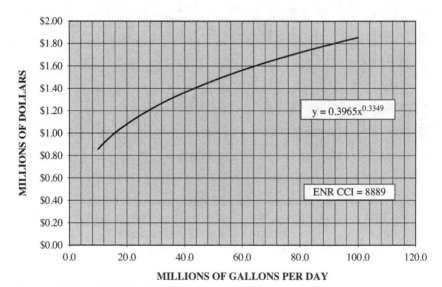

Figure 6.2.2b   O&M Costs for Lime and Soda Ash

6.2.2c Iron Manganese Removal

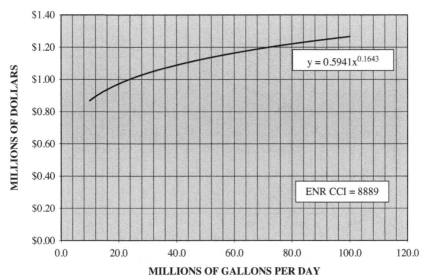

Figure 6.2.2c O&M Costs for Iron Manganese Removal

6.2.3a Micro Membrane Filtration

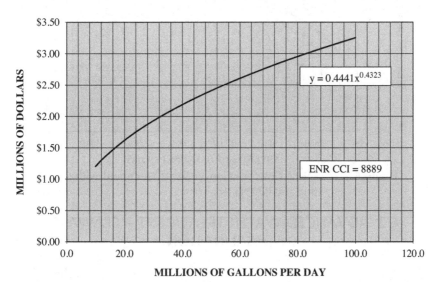

Figure 6.2.3a O&M Costs for Micro Membrane Filtration

6.2.3b   Direct Filtration with Pre-Ozonation

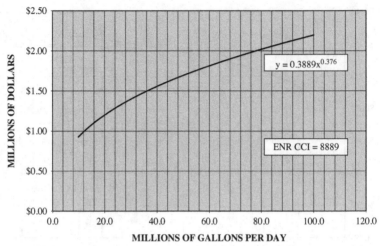

Figure 6.2.3b   O&M Costs for Direction Filtration with Pre-Ozonation

6.2.3c   Conventional Treatment with Ozonation and GAC Filters

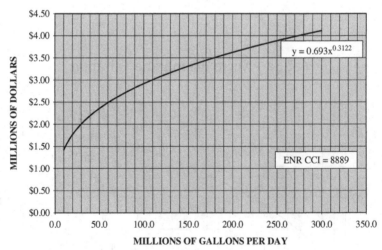

Figure 6.2.3c   O&M Costs for Conventional Treatment with Ozonation and GAC Filters

6.3 ADVANCED WATER TREATMENT – SEAWATER DESALINATION

Cost data for the flow rates shown here vary by treatment type and include a limited amount of data. For desalination, there were from as few

as three to more than ten. The data for ultra- and nano-filtration had significant variation and unsupported source material. These two advanced treatment plant types were combined into one cost curve for O&M.

6.3.1   O&M Costs for Reverse Osmosis Treatment

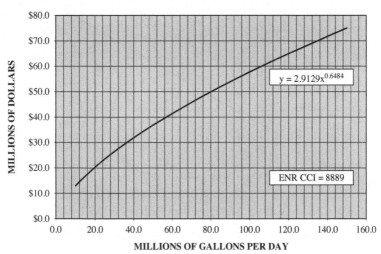

Figure 6.3.1   O&M Costs for Reverse Osmosis

6.3.2   O&M Costs for Multi-Stage Flash

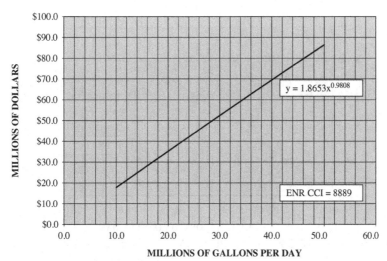

Figure 6.3.2   O&M Costs for Multi-Stage Flash Treatment

6.3.3 O&M Costs for Multiple Effect Distillation (MED) Treatment

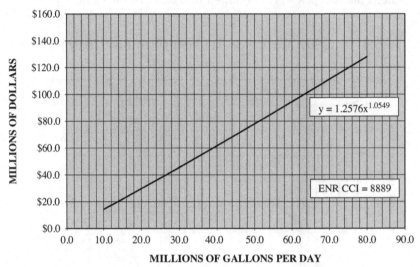

Figure 6.3.3   O&M Costs for Multiple Effect Distillation (MED) Treatment

6.3.4 O&M Costs for Mechanical Vapor Compression (MVC) Treatment

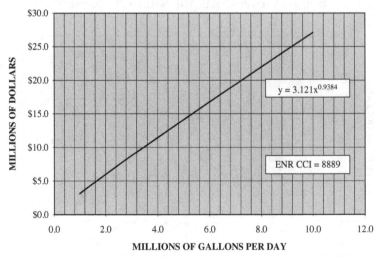

Figure 6.3.4   O&M Costs for Mechanical Vapor Compression (MVC) Treatment

6.3.5   O&M Costs for Ultra-Filtration and Nano-Filtration

The data for ultra-filtration is less detailed than we would prefer and the data on NF is even less robust. The curve below reflects the annual Q&M costs for UF. Since the operating pressure of NF is nearly four times that of UF we expect the Q&M cost for NF to be 2 to 3 times that of the UF Curve Figure 6.3.5.

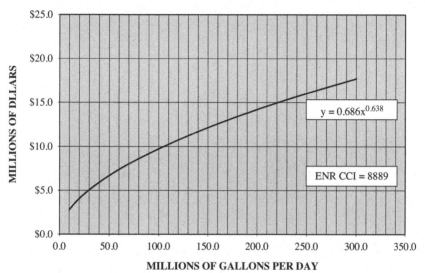

Figure 6.3.5   O&M Cost for UF Filtration

Chapter 7

Future Water Supply, Treatment and Distribution

As previously mentioned, the goal for drinking water quality in the future should be to provide finished water that is safe for human consumption, palatable, and available in quantities necessary to sustain life. Currently, membrane filtration is the primary process to remove suspended particles. This coupled with conventional pretreatment processes like clarification, oxidation, adsorption, and disinfection with UV and chlorination, will be required to adjust the final water quality to maintain the proper mineral content for human health.

Twenty-first century water demand will require treatment schemes that are; high flow rate, effective, reliable, and cost-effective to build and operate, as well as operator-friendly and environmentally friendly. Water treatment in the future will consume more energy than the current conventional treatment processes and delivery schemes. To meet this greater demand and stricter water quality regulations, it will cost much more to design, construct, and operate these facilities. Increased treated water production for human consumption will impose significant impacts on the environment from the disposal of chemical wastes and residue. The economic impact will be more greatly felt in the nations of Africa and the Middle East than in developed countries.

The World Health Organization in *Water a Shared Responsibility, The United Nations World Water Development Report 2*, released in 2006, held that regional economic benefits exceeded costs at all levels of water and sewer treatment intervention. These levels included;

1. Water improvements required to meet the Millennium Development Goals (MDG) of the World Health Organization for water

supply (halving by 2015 the proportion of those without safe drinking water).

2. Water improvements to meet the water MDG for water supply *plus* the MDG for sanitation (halving by 2015 the proportion of those without access to adequate sanitation).
3. Increasing access to improved water and sanitation for everyone.
4. Providing disinfectant at point-of-use over and above increasing access to improved water supply and sanitation.
5. Providing a regulated piped water supply in-house and a sewerage connection with partial sewerage connection for everyone.

The WHO report shows water and sewer treatment benefits dramatically exceeding the annualized capital and O&M costs many times over from a benefit-cost ratio low of 6 to a high of 40. This type of return on capital investment would enhance economic growth and stability, reduce regional conflict, and improve the health of the regions population and their quality of life. Looked at another way not making the investment would reduce the likelihood of economic stability and general health and hygiene, and increase regional conflict, while reducing the economic growth by a like percentage.

There is no single right answer to the question of how to meet the increasing worldwide demand for potable water. Factors such as a reliable water source, geography, the shape and elevation of the surrounding area, and political boundaries all affect the ability to meet the water demands of cities, towns, and villages throughout the world. Well-established urban and suburban areas cannot continue expanding existing water treatment plants and distribution systems indefinitely because of limited land to expand existing facilities.

Wastewater treatment facilities and landfills receiving waste disposal face the same limitations to expansion. Most metropolitan areas of the world have therefore spread out, forming suburbs. In these cases, the best solution is to design and build new (small to midsized) modern water treatment and distribution systems in each region. The availability and treatment requirements are different for each locality: industrial, commercial, and residential zones. Regions with access to the world's oceans could construct seawater desalination plants that met local needs and regulations, with the plants located above the anticipated increase in sea level.

There are multiple design schemes for water treatment plants that could meet future needs under these projected adverse conditions. Three scenarios are presented that could satisfy the mission of public water purveyors. Each design scheme will depend on regional conditions and the regulatory requirements specific these regions.

- Design Scheme 1: Treat and deliver the entire water demand residential consumption, manufacturing, industrial, and commercial services according to strict drinking water quality standards set by regulatory agencies. Although the finished water is of the highest quality, the cost of producing this water may be prohibitive with respect to nonresidential demand.

This design scheme is suitable for most communities in developed countries. Due to the high cost of producing and supplying this water, the consumption of water per capita may have to be limited to as low as 50 gpcd. As an alternative, water consumption could be controlled by metering all service pipelines using a base line cost of water with a high penalty charge for amounts exceeding the limited initial allotment.

- Design Scheme 2: Provide two levels of drinking water quality through the use of two different treatment plants each with its own treatment process train. Approximately 80% of the total water demand would be treated to meet basic regulatory standards using conventional treatment processes and the remaining 20% by an advanced treatment process train. Two sets of water distribution systems would, therefore, be required.

This design scheme may be applied to newly developed communities and to developing nations. The obvious concerns are cross-connections and illegal connections. Although water treated by conventional water treatment processes is relatively safe for human health, people will not become immediately sick or die if the water is accidentally ingested. There are concerns about long-term exposure and cancer, as well as illness due to the contamination of the source water by protozoa.

- Design Scheme 3: Treat all water demand with conventional water treatment processes to satisfy the basic regulatory standards, and provide good-quality drinking water to meet residential demand by installing either a point-of-entry or point-of-use system utilizing a reverse osmosis (RO) process (or equivalent). Alternatively, about 5% of the total demand could be treated with advanced water treatment processes to produce very good water (bottled) for drinking. This bottled water could be delivered by either public or private companies.

This design scheme could be used in remote or isolated regions and developing countries. Such a scheme can provide the minimum daily human needs of 0.5 gpd or even a comfortable 13 gpd for each person and thereby sustain many unserved communities and isolated islands.

Traditionally water supply, wastewater treatment and water reuse have been divided into separate disciplines. In reality they are closely related. Drinking water quality is affected by the efficiency of wastewater treatment. Wastewater treatment plants will discharge treated effluent into large bodies of water or pump it into the ground to become potential source waters. And some manufacturers and industries reclaim their own water multiple times to produce ultra-pure water meet their manufacturing water quality requirements.

It is likely that the demand for potable water will outstrip the supply. Even though water is not considered an "economic good," its presence has always been the precursor to economic growth and its absence the bellwether of economic collapse. The construction cost of water treatment facilities does depend on the scarcity and quality of water supply and the regulatory requirements governing the treatment and delivery of water to the end user.

Appendices

Expanded versions of the tables
in these appendices may be found online at
http://www.wiley.com/WileyCDA/WileyTitle/productCd-0471729973.html

Preface to the Appendices

This appendix contains the tables and graphs used in this manual. These tables and cost curves are shown in the order they appear in the manual and are designed to be used together to arrive at a preliminary construction cost and the associated operations and maintenance for water treatment facilities. They are useful in comparing different design alternatives or plant expansion phases anticipating future development needs. They can also be used in preparation of lifecycle cost to screen multiple alternatives.

A complete set of electronic files created using Microsoft Excel accompany this manual. They are on a single compact disk. There are instructions on the CD in the use of the tables for estimating the construction costs for individual treatment processes, total plant costs and the operation and maintenance costs of the facility. The primary data used to generate the curves and establish equations for calculating the costs are not part of the manual or included on the CD. The files are not protected nor are they guaranteed free of error. They are the result of many adjustments over nearly 30 years of design and construction of water treatment plants.

The user is invited to add their own data to the curves and adjust them for local conditions. But caution should be used in radically changing values either upward or downward to account for anticipated changes in the economy. Most material and labor costs have risen steadily during the past three decades while a few have been relatively flat. The significant increase in development in Asia and the subsequent demand for construction and raw materials has and will continue to drive up the cost for the foreseeable future.

Appendix A

Detailed Treatment Plant Cost Tables

Appendix A1a

Predesign Cost Estimate for a Two-Stage Filtration Treatment Plant Figure 2.3.1a with Capacity of 10 MGD

No.	Process	Applicable Range		Quantity Per Unit Process	Number of Process Units	Process Cost Per Unit	Total Process Cost
		Minimum	Maximum				
1b	Chlorine storage and feed 1-ton cylinder storage	200	10,000	2000	2	$985,458	$1,970,917
6	Liquid Alum Feed - 50% Solution	2	1,000	35	2	$139,493	$278,986
8	Polymer Feed (Cationic)	1	220	20	2	$362,998	$725,996
10	Potassium Permanganate Feed	1	500	40	1	$33,228	$33,228
12	Sodium Hydroxide Feed 50% Solution	1	1,600	400	2	$144,716	$289,432
16	Powdered Activated Carbon	3	6,600	40	1	$250,177	$250,177
18	Rapid Mix G = 600	800	145,000	1750	1	$49,587	$49,587
21	Flocculator G = 50	0.015	7	0.035	2	$297,357	$594,713
22	Flocculator G = 80	0.015	7	0.035	2	$260,163	$520,325
25	Gravity Filter Structure	140	28,000	438	4	$1,131,829	$4,527,318
26	Filtration Media - Stratified Sand	140	28,000	500	4	$26,506	$106,024
27	Filtration Media - Dual Media	140	28,000	400	4	$45,323	$181,294
28	Filtration Media - Mixed Media	140	28,000	400	4	$58,008	$232,032
29	Filter Backwash Pumping	90	1,500	200	2	$186,458	$372,916
31	Air Scour Wash	140	27,000	200	2	$463,853	$927,706

#	Description						
32	Wash Water Surge Basin - (Holding Tank)	9,250	476,000	90000	1	$770,643	$770,643
33	Wash Water Storage Tank - (Waste Holding)	19,800	925,000	200000	1	$216,770	$216,770
34	Clear Water Storage - Below Ground	0.011	8	1.700	1	$1,534,501	$1,534,501
35	Finished Water Pumping - TDH = 100ft - (30.8mts)	1.45	300	4.66	3	$282,487	$847,462
36	Raw Water Pumping	1	200	4	3	$135,079	$405,238
37	Gravity Sludge Thickener	20	150	10	1	$94,864	$94,864
38	Sludge Dewatering lagoons	0.08	40	0.02	3	$4,173	$12,519
42	Administration, Laboratory, and Maintenance Building	1	200	10	1	$280,442	$280,442
	SUB TOTAL PROCESS COSTS						$14,885,033
	YARD PIPING 10%						$1,488,503
	SITEWORK LANDSCAPING 5%						$744,252
	SITE ELECTRICAL & CONTROLS 20%						$2,977,007
	TOTAL CONSTRUCTION COST						$20,094,794
	ENGINEERING, LEGAL & ADMINISTRATIVE COST 35%						$7,033,178
	TOTAL PROJECT COST						$27,127,972

Appendix A1b

Predesign Cost Estimate for a Two-Stage Filtration Treatment Plant Figure 2.3.1a with Capacity of 100 MGD

No.	Process	Applicable Range		Quantity Per Unit Process	Number of Process Units	Process Cost Per Unit	Total Process Cost
		Minimum	Maximum				
1b	Chlorine storage and feed 1-ton cylinder storage	200	10,000	5000	2	$1,807,650	$3,615,300
6	Liquid Alum Feed - 50% Solution	2	1,000	400	2	$454,749	$909,497
8	Polymer Feed (Cationic)	1	220	200	2	$3,398,249	$6,796,498
10	Potassium Permanganate Feed	1	500	400	1	$44,970	$44,970
12	Sodium Hydroxide Feed 50% Solution	1	1,600	1,000	4	$264,300	$1,057,201
16	Powdered Activated Carbon	3	6,600	400	1	$334,106	$334,106
18	Rapid Mix G = 600	800	145,000	17400	1	$128,178	$128,178
21	Flocculator G = 50	0.015	7	0.165	4	$405,535	$1,622,140
22	Flocculator G = 80	0.015	7	0.165	4	$413,130	$1,652,518
25	Gravity Filter Structure	140	28,000	1450	12	$1,843,771	$22,125,254
26	Filtration Media - Stratified Sand	140	28,000	1667	12	$56,126	$673,514
27	Filtration Media - Dual Media	140	28,000	1167	12	$81,600	$979,204
28	Filtration Media - Mixed Media	140	28,000	1167	12	$117,503	$1,410,032
29	Filter Backwash Pumping	90	1,500	600	2	$330,937	$661,875
31	Air Scour Wash	140	27,000	600	2	$497,546	$995,092
32	Wash Water Surge Basin - (Holding Tank)	9,250	476,000	240000	1	$1,608,950	$1,608,950

33	Wash Water Storage Tank - (Waste Holding)	19,800	925,000	250000	2	$261,885	$523,771
34	Clear Water Storage - Below Ground	0.011	8	8.000	2	$6,236,761	$12,473,521
35	Finished Water Pumping - TDH = 100ft - (30.8mts)	1.45	300	31.25	4	$902,691	$3,610,764
36	Raw Water Pumping	1	200	30	4	$525,767	$2,103,069
37	Gravity Sludge Thickener	20	150	10	3	$94,864	$284,592
38	Sludge Dewatering lagoons	0.08	40	1.00	5	$77,538	$387,690
42	Administration, Laboratory, and Maintenance Building	1	200	100	1	$1,001,943	$1,001,943
	SUB TOTAL PROCESS COSTS						$67,151,217
	YARD PIPING 10%						$6,715,122
	SITEWORK LANDSCAPING 5%						$3,357,561
	SITE ELECTRICAL & CONTROLS 20%						$13,430,243
	TOTAL CONSTRUCTION COST						$90,654,143
	ENGINEERING, LEGAL & ADMINISTRATIVE COST 35%						$31,728,950
	TOTAL PROJECT COST						$122,383,092

Appendix A2a

Predesign Cost Estimate for a Direct Filtration Treatment Plant Figure 2.3.1b with Capacity of 10 MGD

No.	Process	Applicable Range		Quantity Per Unit Process	Number of Process Units	Process Cost Per Unit	Total Process Cost
		Minimum	Maximum				
1b	Chlorine storage and feed 1-ton cylinder storage	200	10,000	2000	2	$985,458	$1,970,917
6	Liquid Alum Feed - 50% Solution	2	1,000	50	3	$152,449	$457,346
8	Polymer Feed	1	220	20	2	$362,998	$725,996
10	Potassium Permanganate Feed	1	500	40	1	$33,228	$33,228
12	Sodium Hydroxide Feed 50% Solution	1	1,600	400	2	$144,716	$289,432
16	Powdered Activated Carbon	3	6,600	800	1	$422,051	$422,051
19	Rapid Mix G = 900	800	145,000	1750	2	$56383	$112,766
21	Flocculator G = 50	0.015	7	0.110	2	$359,767	$719,534
25	Gravity Filter Structure	140	28,000	350	4	$1,069,900	$4,279,598
27	Filtration Media - Dual Media	140	28,000	350	4	$42,957	$171,830
29	Filter Backwash Pumping	90	1,500	350	2	$240,638	$481,276
31	Air Scour Wash	140	27,000	350	2	$476,488	$952,976

No.	Description						
32	Wash Water Surge Basin - (Holding Tank)	9,250	476,000	180000	1	$1,296,511	$1,296,511
33	Wash Water Storage Tank	19,800	925,000	200000	1	$216,770	$216,770
34	Clear Water Storage - Below Ground	0.011	8	1.700	1	$1,534,501	$1,534,501
35	Finished Water Pumping - TDH = 100ft	1.45	300	4.66	3	$282,487	$847,462
36	Raw Water Pumping	1	200	4	3	$135,079	$405,238
39	Sand Drying Beds	4,800	400,000	4000	6	$53,724	$322,342
42	Administration, Laboratory, and Maintenance Building	1	200	10	1	$280,442	$280,442

SUB TOTAL PROCESS COSTS	$15,520,215
YARD PIPING 10%	$1,552,021
SITEWORK LANDSCAPING 5%	$776,011
SITE ELECTRICAL & CONTROLS 20%	$3,104,043
TOTAL CONSTRUCTION COST	$20,952,290
ENGINEERING, LEGAL & ADMINISTRATIVE COST 35%	$7,333,302
TOTAL PROJECT COST	$28,285,592

Appendix A2b

Predesign Cost Estimate for a Direct Filtration Treatment Plant Figure 2.3.1b with Capacity of 100 MGD

No.	Process	Applicable Range		Quantity Per Unit Process	Number of Process Units	Process Cost Per Unit	Total Process Cost
		Minimum	Maximum				
1b	Chlorine storage and feed 1-ton cylinder storage	200	10,000	6000	2	$2,039,578	$4,079,155
6	Liquid Alum Feed - 50% Solution	2	1,000	400	3	$454,749	$1,364,246
8	Polymer Feed	1	220	200	2	$3,398,249	$6,796,498
10	Potassium Permanganate Feed	1	500	40	1	$33,228	$33,228
12	Sodium Hydroxide Feed - 50% Solution	1	1,600	4000	2	$862,221	$1,724,443
16	Powdered Activated Carbon	3	6,600	400	2	$334,106	$668,213
19	Rapid Mix G = 900	800	145,000	17400	1	$193,231	$193,231
21	Flocculator G = 50	0.015	7	0.525	4	$705,106	$2,820,425
25	Gravity Filter Structure	140	28,000	1450	12	$1,843,771	$22,125,254
27	Filtration Media - Dual Media	140	28,000	1450	12	$95,007	$1,140,085
29	Filter Backwash Pumping	90	1,500	600	2	$330,937	$661,875
31	Air Scour Wash	140	27,000	1450	2	$569,143	$1,138,286

No.	Description						
32	Wash Water Surge Basin - (Holding Tank)	9,250	476,000	240000	1	$1,608,950	$1,608,950
33	Wash Water Storage Tank - (Waste Holding)	19,800	925,000	250000	2	$261,885	$523,771
34	Clear Water Storage - Below Ground	0.011	8	8.000	2	$6,236,761	$12,473,521
35	Finished Water Pumping - TDH = 100ft	1.45	300	31.25	4	$902,691	$3,610,764
36	Raw Water Pumping	1	200	30	4	$525,767	$2,103,069
37	Gravity Sludge Thickener	20	150	10	3	$94,864	$284,592
40	Filter Press	30	6,600	1000	1	$1,802,536	$1,802,536
42	Administration, Laboratory, and Maintenance Building	1	200	100	1	$1,001,943	$1,001,943

SUB TOTAL PROCESS COSTS		$66,154,084
YARD PIPING 10%		$6,615,408
SITEWORK LANDSCAPING 5%		$3,307,704
SITE ELECTRICAL & CONTROLS 20%		$13,230,817
TOTAL CONSTRUCTION COST		$89,308,013
ENGINEERING, LEGAL & ADMINISTRATIVE COST 35%		$31,257,804
TOTAL PROJECT COST		$120,565,817

Appendix A3a

Predesign Cost Estimate for a Conventional Treatment Plant Figure 2.3.1c with Capacity of 10 MGD

No.	Process	Applicable Range		Quantity Per Unit Process	Number of Process Units	Process Cost Per Unit	Total Process Cost
		Minimum	Maximum				
1b	Chlorine storage and feed 1-ton cylinder storage	200	10,000	2000	2	$985,458	$1,970,917
6	Liquid Alum Feed - 50% Solution	2	1,000	150	2	$238,820	$477,640
8	Polymer Feed	1	220	20	2	$362,998	$725,996
10	Potassium Permanganate Feed	1	500	40	1	$33,228	$33,228
12	Sodium Hydroxide Feed 50% Solution	1	1,600	400	2	$144,716	$289,432
15	Aqua Ammonia Feed	240	5,080	30	2	$27,189	$54,378
16	Powdered Activated Carbon	3	6,600	40	2	$250,177	$500,355
19	Rapid Mix G = 900	800	145,000	1750	2	$56,383	$112,766
20	Flocculator G = 20	0.015	7	0.035	2	$301,986	$603,971
21	Flocculator G = 50	0.015	7	0.035	2	$297,357	$594,713
22	Flocculator G = 80	0.015	7	0.035	2	$260,163	$520,325
24	Rectangular Clarier	5,000	150,000	5000	2	$252,180	$504,359
25	Gravity Filter Structure	140	28,000	350	4	$1,069,900	$4,279,598
27	Filtration Media - Dual Media	140	28,000	350	4	$42,957	$171,830

132

29	Filter Backwash Pumping	90	1,500	350	2	$240,638	$481,276
31	Air Scour Wash	140	27,000	350	2	$476,488	$952,976
32	Wash Water Surge Basin - (Holding Tank)	9,250	476,000	90000	1	$770,643	$770,643
33	Wash Water Storage Tank	19,800	925,000	200000	1	$216,770	$216,770
34	Clear Water Storage - Below Ground	0.011	8	1.700	1	$1,534,501	$1,534,501
35	Finished Water Pumping - TDH = 100ft	1.45	300	4.66	3	$282,487	$847,462
36	Raw Water Pumping	1	200	4	3	$135,079	$405,238
37	Gravity Sludge Thickener	20	150	20	1	$234,394	$234,394
39	Sand Drying Beds	4,800	400,000	5333	6	$69,103	$414,620
42	Administration, Laboratory, and Maintenance Building	1	200	10	1	$280,442	$280,442

SUB TOTAL PROCESS COSTS	$16,977,829
YARD PIPING 10%	$1,697,783
SITEWORK LANDSCAPING 5%	$848,891
SITE ELECTRICAL & CONTROLS 20%	$3,395,566
TOTAL CONSTRUCTION COST	$22,920,069
ENGINEERING, LEGAL & ADMINISTRATIVE COST 35%	$8,022,024
TOTAL PROJECT COST	$30,942,093

Appendix A3b

Predesign Cost Estimate for a Conventional Treatment Plant Figure 2.3.1c with Capacity of 100 MGD

No.	Process	Applicable Range		Quantity Per Unit Process	Number of Process Units	Process Cost Per Unit	Total Process Cost
		Minimum	Maximum				
1b	Chlorine storage and feed 1-ton cylinder storage	200	10,000	4000	3	$1,559,374	$4,678,121
6	Liquid Alum Feed - 50% Solution	2	1,000	1667	3	$1,548,787	$4,646,360
8	Polymer Feed	1	220	200	2	$3,398,249	$6,796,498
10	Potassium Permanganate Feed	1	500	40	1	$33,228	$33,228
12	Sodium Hydroxide Feed - 50% Solution	1	1,600	4000	2	$862,221	$1,724,443
16	Powdered Activated Carbon	3	6,600	400	2	$334,106	$668,213
19	Rapid Mix G = 900	800	145,000	17400	1	$193,231	$193,231
20	Flocculator G = 20	0.015	7	0.175	4	$399,841	$1,599,365
21	Flocculator G = 50	0.015	7	0.175	4	$413,856	$1,655,426
22	Flocculator G = 80	0.015	7	0.175	4	$424,896	$1,699,585
24	Rectangular Clarifier	5,000	150,000	16667	4	$369,905	$1,479,619
25	Gravity Filter Structure	140	28,000	1450	12	$1,843,771	$22,125,254
27	Filtration Media - Dual Media	140	28,000	1450	12	$95,007	$1,140,085
29	Filter Backwash Pumping	90	1,500	600	2	$330,937	$661,875
31	Air Scour Wash	140	27,000	1450	2	$569,143	$1,138,286

#	Description						
32	Wash Water Surge Basin - (Holding Tank)	9,250	476,000	240000	1	$1,608,950	$1,608,950
33	Wash Water Storage Tank	19,800	925,000	250000	2	$261,885	$523,771
34	Clear Water Storage - Below Ground	0.011	8	8.000	2	$6,236,761	$12,473,521
35	Finished Water Pumping - TDH = 100ft	1.45	300	31.25	4	$902,691	$3,610,764
36	Raw Water Pumping	1	200	30	4	$525,767	$2,103,069
37	Gravity Sludge Thickener	20	150	17	3	$184,763	$554,288
38	Sludge Dewatering lagoons	0.08	40	1.67	6	$111,647	$669,883
40	Filter Press	30	6,600	1000	1	$1,802,536	$1,802,536
42	Administration, Laboratory, and Maintenance Building	1	200	100	1	$1,001,943	$1,001,943

SUB TOTAL PROCESS COSTS	$74,588,312
YARD PIPING 10%	$7,458,831
SITEWORK LANDSCAPING 5%	$3,729,416
SITE ELECTRICAL & CONTROLS 20%	$14,917,662
TOTAL CONSTRUCTION COST	$100,694,222
ENGINEERING, LEGAL & ADMINISTRATIVE COST 35%	$35,242,978
TOTAL PROJECT COST	$135,937,199

Appendix A4a

Predesign Cost Estimate for a Dissolved Air Flotation Treatment Plant Figure 2.4.1a with Capacity of 10 MGD

No.	Process	Range Minimum	Maximum	Quantity Per Unit Process	Number of Process Units	Process Cost Per Unit	Total Process Cost
1b	Chlorine storage and feed 1-ton cylinder storage	200	10,000	2000	2	$985,458	$1,970,917
6	Liquid Alum Feed - 50% Solution	2	1,000	75	2	$174,041	$348,083
8	Polymer Feed	1	220	20	1	$362,998	$362,998
10	Potassium Permanganate Feed	1	500	40	1	$33,228	$33,228
12	Sodium Hydroxide Feed 50% Solution	1	1,600	400	2	$144,716	$289,432
19	Rapid Mix G = 900	800	145,000	1750	1	$56,383	$56,383
21	Flocculator G = 50	0.015	7	0.053	2	$312,335	$624,670
22	Flocculator G = 80	0.015	7	0.053	2	$281,343	$562,685
23	Circular Clarifier (10 ft walls)	650	32,300	35	2	$52,324	$104,648
24	Rectangular Clarifier	5,000	150,000	5000	2	$252,180	$504,359
25	Gravity Filter Structure	140	28,000	450	4	$1,140,671	$4,562,685
27	Filtration Media - Dual Media	140	28,000	450	4	$47,689	$190,757
29	Filter Backwash Pumping	90	1,500	450	2	$276,758	$553,515
31	Air Scour Wash	140	27,000	450	2	$484,911	$969,822

32	Wash Water Surge Basin - (Holding Tank)	9,250	476,000	90000	1	$770,643	$770,643
33	Wash Water Storage Tank	19,800	925,000	200000	1	$216,770	$216,770
34	Clear Water Storage - Below Ground	0.011	8	1.700	1	$1,534,501	$1,534,501
35	Finished Water Pumping - TDH = 100ft	1.45	300	4.66	3	$282,487	$847,462
36	Raw Water Pumping	1	200	4	3	$135,079	$405,238
39	Sand Drying Beds	4,800	400,000	4000	8	$53,724	$429,789
42	Administration, Laboratory, and Maintenance Building	1	200	10	1	$280,442	$280,442

SUB TOTAL PROCESS COSTS	$15,619,028
YARD PIPING 10%	$1,561,903
SITEWORK LANDSCAPING 5%	$780,951
SITE ELECTRICAL & CONTROLS 20%	$3,123,806
TOTAL CONSTRUCTION COST	$21,085,687
ENGINEERING, LEGAL & ADMINISTRATIVE COST 35%	$7,379,991
TOTAL PROJECT COST	$28,465,678

Appendix A4b

Predesign Cost Estimate for a Dissolved Air Flotation Treatment Plant Figure 2.4.1a with Capacity of 100 MGD

No.	Process	Applicable Range		Quantity Per Unit Process	Number of Process Units	Process Cost Per Unit	Total Process Cost
		Minimum	Maximum				
1b	Chlorine storage and feed 1-ton cylinder storage	200	10,000	5000	2	$1,807,650	$3,615,300
6	Liquid Alum Feed - 50% Solution	2	1,000	467	3	$512,330	$1,536,989
8	Polymer Feed	1	220	175	2	$2,976,686	$5,953,373
10	Potassium Permanganate Feed	1	500	400	1	$44,970	$44,970
12	Sodium Hydroxide Feed - 50% Solution	1	1,600	4000	2	$862,221	$1,724,443
19	Rapid Mix G = 900	800	145,000	17400	2	$193,231	$386,462
22	Flocculator G = 80	0.015	7	0.175	4	$424,896	$1,699,585
24	Rectangular Clarifier	5,000	150,000	10000	4	$314,411	$1,257,644
25	Gravity Filter Structure	140	28,000	600	24	$1,246,672	$29,920,119
27	Filtration Media - Dual Media	140	28,000	600	24	$54,787	$1,314,886
29	Filter Backwash Pumping	90	1,500	600	4	$330,937	$1,323,749
31	Air Scour Wash	140	27,000	600	4	$497,546	$1,990,183

No.	Description						
32	Wash Water Surge Basin - (Holding Tank)	9,250	476,000	240000	1	$1,608,950	$1,608,950
33	Wash Water Storage Tank	19,800	925,000	500000	1	$471,166	$471,166
34	Clear Water Storage - Below Ground	0.011	8	8.000	2	$6,236,761	$12,473,521
35	Finished Water Pumping - TDH = 100ft	1.45	300	31.25	4	$902,691	$3,610,764
36	Raw Water Pumping	1	200	30	4	$525,767	$2,103,069
39	Sand Drying Beds	4,800	400,000	26667	12	$282,597	$3,391,167
42	Administration, Laboratory, and Maintenance Building	1	200	100	1	$1,001,943	$1,001,943
	SUB TOTAL PROCESS COSTS						$75,428,284
	YARD PIPING 10%						$7,542,828
	SITEWORK LANDSCAPING 5%						$3,771,414
	SITE ELECTRICAL & CONTROLS 20%						$15,085,657
	TOTAL CONSTRUCTION COST						$101,828,183
	ENGINEERING, LEGAL & ADMINISTRATIVE COST 35%						$35,639,864
	TOTAL PROJECT COST						$137,468,047

Appendix A5a

Predesign Cost Estimate for a Lime Soda Ash Treatment Plant Figure 2.4.1b with Capacity of 10 MGD

No.	Process	Applicable Range		Quantity Per Unit Process	Number of Process Units	Process Cost Per Unit	Total Process Cost
		Minimum	Maximum				
1b	Chlorine storage and feed 1-ton cylinder storage	200	10,000	2000	2	$985,458	$1,970,917
9	Lime Feed	10	10,000	1600	2	$1,246,221	$2,492,441
19	Rapid Mix G = 900	800	145,000	1750	1	$56,383	$56,383
20	Flocculator G = 20	0.015	7	0.035	2	$301,986	$603,971
21	Flocculator G = 50	0.015	7	0.035	2	$297,357	$594,713
22	Flocculator G = 80	0.015	7	0.035	2	$260,163	$520,325
23	Circular Clarifier	650	32,300	35	2	$52,324	$104,648
24	Rectangular Clarifier	5,000	150,000	5000	4	$252,180	$1,008,718
25	Gravity Filter Structure	140	28,000	450	4	$1,140,671	$4,562,685
27	Filtration Media - Dual Media	140	28,000	450	4	$47,689	$190,757
29	Filter Backwash Pumping	90	1,500	200	2	$186,458	$372,916
30	Surface Wash System	140	27,000	200	2	$99,941	$199,881
31	Air Scour Wash	140	27,000	200	2	$463,853	$927,706
32	Wash Water Surge Basin - (Holding Tank)	9,250	476,000	90000	1	$770,643	$770,643

33	Wash Water Storage Tank	19,800	925,000	200000	1	$216,770	$216,770
34	Clear Water Storage - Below Ground	0.011	8	1.700	1	$1,534,501	$1,534,501
35	Finished Water Pumping - TDH = 100ft	1.45	300	4.66	3	$282,487	$847,462
36	Raw Water Pumping	1	200	4	3	$135,079	$405,238
37	Gravity Sludge Thickener	20	150	35	2	$486,531	$973,061
39	Sand Drying Beds	4,800	400,000	1875	8	$27,683	$221,460
42	Administration, Laboratory, and Maintenance Building	1	200	10	1	$280,442	$280,442

SUB TOTAL PROCESS COSTS			$18,855,641
YARD PIPING 10%			$1,885,564
SITEWORK LANDSCAPING 5%			$942,782
SITE ELECTRICAL & CONTROLS 20%			$3,771,128
TOTAL CONSTRUCTION COST			$25,455,115
ENGINEERING, LEGAL & ADMINISTRATIVE COST 35%			$8,909,290
TOTAL PROJECT COST			$34,364,405

Note: CO_2 gas is used for decarbonation as an alternative to H_2SO_4 see Figure 2.4.1b. The application of CO_2 gas directly injected into the effluent of the Solids Contact Clarifier is acceptable and only impacts the O&M costs.

Appendix A5b

Predesign Cost Estimate for a Lime Soda Ash Treatment Plant Figure 2.4.1b with Capacity of 100 MGD

No.	Process	Applicable Range Minimum	Maximum	Quantity Per Unit Process	Number of Process Units	Process Cost Per Unit	Total Process Cost
1b	Chlorine storage and feed 1-ton cylinder storage	200	10,000	5000	2	$1,807,650	$3,615,300
9	Lime Feed	10	10,000	160000	2	$18,870,999	$37,741,998
19	Rapid Mix G = 900	800	145,000	17400	2	$193,231	$386,462
20	Flocculator G = 20	0.015	7	0.165	4	$392,852	$1,571,406
21	Flocculator G = 50	0.015	7	0.165	4	$405,535	$1,622,140
22	Flocculator G = 80	0.015	7	0.165	4	$413,130	$1,652,518
23	Circular Clarifier	650	32,300	130	12	$117,622	$1,411,459
24	Rectangular Clarifier	5,000	150,000	16667	12	$369,905	$4,438,856
25	Gravity Filter Structure	140	28,000	1450	12	$1,843,771	$22,125,254
27	Filtration Media - Dual Media	140	28,000	1450	12	$95,007	$1,140,085
29	Filter Backwash Pumping	90	1,500	600	2	$330,937	$661,875
30	Surface Wash System	140	27,000	600	2	$128,863	$257,725
31	Air Scour Wash	140	27,000	600	2	$497,546	$995,092
32	Wash Water Surge Basin - (Holding Tank)	9,250	476,000	240000	1	$1,608,950	$1,608,950

33	Wash Water Storage Tank	19,800	925,000	250000	2	$261,885	$523,771
34	Clear Water Storage - Below Ground	0.011	8	8.000	2	$6,236,761	$12,473,521
35	Finished Water Pumping - TDH = 100ft	1.45	300	31.25	4	$902,691	$3,610,764
36	Raw Water Pumping	1	200	30	4	$525,767	$2,103,069
37	Gravity Sludge Thickener	20	150	20	4	$234,394	$937,574
39	Sand Drying Beds	4,800	400,000	13333	12	$154,077	$1,848,919
42	Administration, Laboratory, and Maintenance Building	1	200	100	1	$1,001,943	$1,001,943

SUB TOTAL PROCESS COSTS	$101,728,681
YARD PIPING 10%	$10,172,868
SITEWORK LANDSCAPING 5%	$5,086,434
SITE ELECTRICAL & CONTROLS 20%	$20,345,736
TOTAL CONSTRUCTION COST	$137,333,720
ENGINEERING, LEGAL & ADMINISTRATIVE COST 35%	$48,066,802
TOTAL PROJECT COST	$185,400,522

Note: CO_2 gas is used for decarbonation as an alternative to H_2SO_4, see Figure 2.4.1b. The application of CO_2 gas directly injected into the effluent of the Solids Contact Clarifier is acceptable and only impacts the O&M.

Appendix A6a

Predesign Cost Estimate Iron Manganese Removal Treatment Plant Figure 2.4.1c with Capacity of 10 MGD

No.	Process	Applicable Range		Quantity Per Unit Process	Number of Process Units	Process Cost Per Unit	Total Process Cost
		Minimum	Maximum				
1b	Chlorine storage and feed 1-ton cylinder storage	200	10,000	2000	2	$985,458	$1,970,917
6	Liquid Alum Feed - 50% Solution	2	1,000	150	2	$238,820	$477,640
9	Lime Feed	10	10,000	400	2	$549,945	$1,099,890
10	Potassium Permanganate Feed	1	500	125	2	$36,000	$72,001
12	Sodium Hydroxide Feed - 50% Solution	1	1,600	400	2	$144,716	$289,432
19	Rapid Mix G = 900	800	145,000	1750	1	$56,383	$56,383
20	Flocculator G = 20	0.015	7	0.035	2	$301,986	$603,971
21	Flocculator G = 50	0.015	7	0.035	2	$297,357	$594,713
22	Flocculator G = 80	0.015	7	0.035	2	$260,163	$520,325
24	Rectangular Clarifier	5,000	150,000	5000	2	$252,180	$504,359
25	Gravity Filter Structure	140	28,000	450	4	$1,140,671	$4,562,685
27	Filtration Media - Dual Media	140	28,000	450	4	$47,689	$190,757
29	Filter Backwash Pumping	90	1,500	450	4	$276,758	$1,107,031

31	Air Scour Wash	140	27,000	450	4	$484,911	$1,939,644
32	Wash Water Surge Basin - (Holding Tank)	9,250	476,000	90000	1	$770,643	$770,643
33	Wash Water Storage Tank	19,800	925,000	200000	1	$216,770	$216,770
34	Clear Water Storage - Below Ground	0.011	8	1.700	1	$1,534,501	$1,534,501
35	Finished Water Pumping - TDH = 100ft	1.45	300	4.66	3	$282,487	$847,462
36	Raw Water Pumping	1	200	4	3	$135,079	$405,238
37	Gravity Sludge Thickener	20	150	35	2	$486,531	$973,061
39	Sand Drying Beds	4,800	400,000	375	8	$6,769	$54,154
42	Administration, Laboratory, and Maintenance Building	1	200	10	1	$280,442	$280,442

SUB TOTAL PROCESS COSTS	$19,072,019
YARD PIPING 10%	$1,907,202
SITEWORK LANDSCAPING 5%	$953,601
SITE ELECTRICAL & CONTROLS 20%	$3,814,404
TOTAL CONSTRUCTION COST	$25,747,225
ENGINEERING, LEGAL & ADMINISTRATIVE COST 35%	$9,011,529
TOTAL PROJECT COST	$34,758,754

Appendix A6b

Predesign Cost Estimate for a Iron Manganese Removal Treatment Plant Figure 2.4.1c with Capacity of 100 MGD

No.	Process	Applicable Range		Quantity Per Unit Process	Number of Process Units	Process Cost Per Unit	Total Process Cost
		Minimum	Maximum				
1b	Chlorine storage and feed 1-ton cylinder storage	200	10,000	4000	3	$1,559,374	$4,678,121
6	Liquid Alum - 50% Solution	2	1000	833	3	$829,025	$2,487,074
8	Polymer Feed	1	220	80	2	$1,374,748	$2,749,497
10	Potassium Permanganate Feed	1	500	4000	2	$162,396	$324,793
11	Sulfuric Acid Feed 93% Solution	11	5,300	1250	2	$82,783	$165,565
12	Sodium Hydroxide Feed - 50% Solution	1	1,600	800	4	$224,439	$897,755
19	Rapid Mix G = 900	800	145,000	17400	1	$193,231	$193,231
20	Flocculator G = 20	0.015	7	0.165	4	$392,852	$1,571,406
21	Flocculator G = 50	0.015	7	0.165	4	$405,535	$1,622,140
22	Flocculator G = 80	0.015	7	0.165	4	$413,130	$1,652,518
24	Rectangular Clarifier	5,000	150,000	16667	6	$369,905	$2,219,428
25	Gravity Filter Structure	140	28,000	1450	12	$1,843,771	$22,125,254
27	Filtration Media - Dual Media	140	28,000	1450	12	$95,007	$1,140,085
29	Filter Backwash Pumping	90	1,500	1450	2	$637,956	$1,275,911

#							
31	Air Scour Wash	140	27,000	1450	2	$569,143	$1,138,286
32	Wash Water Surge Basin - (Holding Tank)	9,250	476,000	240000	1	$1,608,950	$1,608,950
33	Wash Water Storage Tank - (Waste Holding)	19,800	925,000	250000	2	$261,885	$523,771
34	Clear Water Storage - Below Ground	0.011	8	8.000	2	$6,236,761	$12,473,521
35	Finished Water Pumping - TDH = 100ft	1.45	300	31.25	4	$902,691	$3,610,764
36	Raw Water Pumping	1	200	30	4	$525,767	$2,103,069
37	Gravity Sludge Thickener	20	150	17	3	$184,763	$554,288
39	Sand Drying Beds	4,800	400,000	26667	12	$282,597	$3,391,167
42	Administration, Laboratory, and Maintenance Building	1	200	100	1	$1,001,943	$1,001,943

SUB TOTAL PROCESS COSTS	$69,508,537
YARD PIPING 10%	$6,950,854
SITEWORK LANDSCAPING 5%	$3,475,427
SITE ELECTRICAL & CONTROLS 20%	$13,901,707
TOTAL CONSTRUCTION COST	$93,836,525
ENGINEERING, LEGAL & ADMINISTRATIVE COST 35%	$32,842,784
TOTAL PROJECT COST	$126,679,309

Appendix A7a

Predesign Cost Estimate for a Micro Membrane Treatment Plant Figure 2.4.2a with Capacity of 10 MGD

No.	Process	Applicable Range Minimum	Applicable Range Maximum	Quantity Per Unit Process	Number of Process Units	Process Cost Per Unit	Total Process Cost
1b	Chlorine storage and feed 1-ton cylinder storage	200	10,000	2000	2	$985,458	$1,970,917
10	Potassium Permanganate Feed	1	500	400	2	$44,970	$89,941
11	Sulfuric Acid Feed - 93% Solution	11	5,300	125	2	$37,508	$75,016
12	Sodium Hydroxide Feed - 50% Solution	1	1,600	400	2	$144,716	$289,432
16	Powdered Activated Carbon	3	6,600	10	1	$242,979	$242,979
18	Rapid Mix G = 600	800	145,000	1750	1	$49,587	$49,587
19	Rapid Mix G = 900	800	145,000	1750	1	$56,383	$56,383
25	Proprietary Membrane Unit	140	28,000				$20,000,000

29	Proprietary Chemical Wash	90	1,500				$250,000
34	Clear Water Storage - Below Ground	0.011	8	1.700	1	$1,534,501	$1,534,501
35	Finished Water Pumping - TDH = 100ft	1.45	300	4.66	3	$282,487	$847,462
36	Raw Water Pumping	1	200	4	3	$135,079	$405,238
42	Administration, Laboratory, and Maintenance Building	1	200	10	2	$280,442	$560,884

SUB TOTAL PROCESS COSTS	$26,372,338
YARD PIPING 10%	$2,637,234
SITEWORK LANDSCAPING 5%	$1,318,617
SITE ELECTRICAL & CONTROLS 20%	$5,274,468
TOTAL CONSTRUCTION COST	$35,602,656
ENGINEERING, LEGAL & ADMINISTRATIVE COST 35%	$12,460,930
TOTAL PROJECT COST	$48,063,586

Appendix A7b

Predesign Cost Estimate for a Micro Membrane Treatment Plant Figure 2.4.2a with Capacity of 10 MGD

No.	Process	Applicable Range Minimum	Applicable Range Maximum	Quantity Per Unit Process	Number of Process Units	Process Cost Per Unit	Total Process Cost
1b	Chlorine storage and feed 1-ton cylinder storage	200	10,000	5000	2	$1,807,650	$3,615,300
10	Potassium Permanganate Feed	1	500	4000	2	$162,396	$324,793
11	Sulfuric Acid Feed - 93% Solution	11	5,300	1250	2	$82,783	$165,565
12	Sodium Hydroxide Feed - 50% Solution	1	1,600	4000	2	$862,221	$1,724,443
18	Rapid Mix G = 600	800	145,000	17400	1	$128,178	$128,178
19	Rapid Mix G = 900	800	145,000	17400	1	$193,231	$193,231
25	Proprietary Membrane Unit	140	28,000				$150,000,000

No.	Description						
27	Proprietary Chemical Wash	140	28,000				$1,000,000
34	Clear Water Storage - Below Ground	0.011	8	8.000	2	$6,236,761	$12,473,521
35	Finished Water Pumping - TDH = 100ft	1.45	300	31.25	4	$902,691	$3,610,764
36	Raw Water Pumping	1	200	30	4	$525,767	$2,103,069
42	Administration, Laboratory, and Maintenance Building	1	200	100	3	$1,001,943	$3,005,830

SUB TOTAL PROCESS COSTS	$178,344,693
YARD PIPING 10%	$17,834,469
SITEWORK LANDSCAPING 5%	$8,917,235
SITE ELECTRICAL & CONTROLS 20%	$35,668,939
TOTAL CONSTRUCTION COST	$240,765,336
ENGINEERING, LEGAL & ADMINISTRATIVE COST 35%	$84,267,868
TOTAL PROJECT COST	$325,033,203

Appendix A8a

Predesign Cost Estimate for a Direct Filtration with Pre-Ozone Treatment Plant Figure 2.4.2b with Capacity of 10 MGD

No.	Process	Applicable Range		Quantity Per Unit Process	Number of Process Units	Process Cost Per Unit	Total Process Cost
		Minimum	Maximum				
1b	Chlorine storage and feed 1-ton cylinder storage	200	10,000	2000	2	$985,458	$1,970,917
4	Ozone Generation	10	3,500	250	2	$1,366,570	$2,733,140
5	Ozone Contact Chamber	1,060	423,000	35000	2	$93,141	$186,281
6	Liquid Alum Feed - 50% Solution	2	1,000	50	3	$152,449	$457,346
8	Polymer Feed	1	220	20	2	$362,998	$725,996
12	Sodium Hydroxide Feed - 50% Solution	1	1,600	400	2	$144,716	$289,432
19	Rapid Mix G = 900	800	145,000	1750	2	$56,383	$112,766
21	Flocculator G = 50	0.015	7	0.110	2	$359,767	$719,534
25	Gravity Filter Structure	140	28,000	350	4	$1,069,900	$4,279,598
27	Filtration Media - Dual Media	140	28,000	350	4	$42,957	$171,830
29	Filter Backwash Pumping	90	1,500	350	2	$240,638	$481,276
31	Air Scour Wash	140	27,000	350	2	$476,488	$952,976

No.	Description						
32	Wash Water Surge Basin - (Holding Tank)	9,250	476,000	180000	1	$1,296,511	$1,296,511
33	Wash Water Storage Tank - (Waste Wash Water)	19,800	925,000	200000	1	$216,770	$216,770
34	Clear Water Storage - Below Ground	0.011	8	1.700	1	$1,534,501	$1,534,501
35	Finished Water Pumping - TDH = 100ft	1.45	300	4.66	3	$282,487	$847,462
36	Raw Water Pumping	1	200	4	3	$135,079	$405,238
39	Sand Drying Beds	4,800	400,000	4000	6	$53,724	$322,342
42	Administration, Laboratory, and Maintenance Building	1	200	10	1	$280,442	$280,442

SUB TOTAL PROCESS COSTS	$17,984,357
YARD PIPING 10%	$1,798,436
SITEWORK LANDSCAPING 5%	$899,218
SITE ELECTRICAL & CONTROLS 20%	$3,596,871
TOTAL CONSTRUCTION COST	$24,278,881
ENGINEERING, LEGAL & ADMINISTRATIVE COST 35%	$8,497,608
TOTAL PROJECT COST	$32,776,490

Appendix A8b

Predesign Cost Estimate for a Direct Filtration Treatment Plant with Pre-Ozone Figure 2.4.2b with Capacity of 100 MGD

No.	Process	Applicable Range		Quantity Per Unit Process	Number of Process Units	Process Cost Per Unit	Total Process Cost
		Minimum	Maximum				
1b	Chlorine storage and feed 1-ton cylinder storage	200	10,000	6000	2	$2,039,578	$4,079,155
4	Ozone Generation	10	3,500	2500	2	$6,069,205	$12,138,410
5	Ozone Contact Chamber	1,060	423,000	173750	4	$261,463	$1,045,851
7	Dry Alum Feed	10	5,070	267	3	$160,258	$480,774
8	Polymer Feed	1	220	117	3	$1,993,040	$5,979,121
9	Lime Feed	10	10,000	4000	2	$2,140,027	$4,280,054
19	Rapid Mix G = 900	800	145,000	17400	1	$193,231	$193,231
21	Flocculator G = 50	0.015	7	0.175	4	$413,856	$1,655,426
22	Flocculator G = 80	0.015	7	0.175	4	$424,896	$1,699,585
25	Gravity Filter Structure	140	28,000	1450	12	$1,843,771	$22,125,254
27	Filtration Media - Dual Media	140	28,000	1167	12	$81,600	$979,204
29	Filter Backwash Pumping	90	1,500	600	2	$330,937	$661,875
31	Air Scour Wash	140	27,000	600	2	$497,546	$995,092
32	Wash Water Surge Basin - (Holding Tank)	9,250	476,000	240000	1	$1,608,950	$1,608,950

No.	Description						
33	Wash Water Storage Tank - (Waste Wash Water)	19,800	925,000	250000	2	$261,885	$523,771
34	Clear Water Storage - Below Ground	0.011	8	8.000	2	$6,236,761	$12,473,521
35	Finished Water Pumping - TDH = 100ft	1.45	300	31.25	4	$902,691	$3,610,764
36	Raw Water Pumping	1	200	30	4	$525,767	$2,103,069
37	Gravity Sludge Thickener	20	150	10	3	$94,864	$284,592
38	Sludge Dewatering lagoons	0.08	40	1.00	5	$77,538	$387,690
39	Sand Drying Beds	4,800	400,000	15000	10	$170,805	$1,708,048
42	Administration, Laboratory, and Maintenance Building	1	200	100	1	$1,001,943	$1,001,943

SUB TOTAL PROCESS COSTS	$80,015,380
YARD PIPING 10%	$8,001,538
SITEWORK LANDSCAPING 5%	$4,000,769
SITE ELECTRICAL & CONTROLS 20%	$16,003,076
TOTAL CONSTRUCTION COST	$108,020,763
ENGINEERING, LEGAL & ADMINISTRATIVE COST 35%	$37,807,267
TOTAL PROJECT COST	$145,828,030

Appendix A9a

Predesign Cost Estimate for a Iron Manganese Removal Treatment Plant Figure 2.4.2c with Capacity of 10 MGD

No.	Process	Applicable Range		Quantity Per Unit Process	Number of Process Units	Process Cost Per Unit	Total Process Cost
		Minimum	Maximum				
1b	Chlorine storage and feed 1-ton cylinder storage	200	10,000	2000	2	$985,458	$1,970,917
4	Ozone Generation	10	3,500	250	2	$1,366,570	$2,733,140
5	Ozone Contact Chamber	1,060	423,000	35000	2	$93,141	$186,281
6	Liquid Alum Feed - 50% Solution	2	1,000	150	2	$238,820	$477,640
8	Polymer Feed	1	220	20	2	$362,998	$725,996
12	Sodium Hydroxide Feed - 50% Solution	1	1,600	400	2	$144,716	$289,432
19	Rapid Mix G = 900	800	145,000	1750	2	$56,383	$112,766
20	Flocculator G = 20	0.015	7	0.035	2	$301,986	$603,971
21	Flocculator G = 50	0.015	7	0.035	2	$297,357	$594,713
22	Flocculator G = 80	0.015	7	0.035	2	$260,163	$520,325
24	Rectangular Clarifier	5,000	150,000	5000	2	$252,180	$504,359
25	Gravity Filter Structure	140	28,000	450	4	$1,140,671	$4,562,685
27	Filtration Media - Dual Media	140	28,000	450	4	$47,689	$190,757
29	Filter Backwash Pumping	90	1,500	450	2	$276,758	$553,515
31	Air Scour Wash	140	27,000	450	2	$484,911	$969,822

#	Description						
32	GAC Filtration/Adsorption Bed	140	28,000	450	5	$1,140,671	$5,703,357
33	GAC Media	140	28,000	450	5	$381,514	$1,907,571
34	Wash Water Surge Basin - (Holding Tank)	9,250	476,000	90000	1	$770,643	$770,643
35	Wash Water Storage Tank - (Waste Wash Water)	19,800	925,000	200000	1	$216,770	$216,770
36	Clear Water Storage - Below Ground	0.011	8	1.700	1	$1,534,501	$1,534,501
37	Finished Water Pumping TDH = 100ft	1.45	300	4.66	3	$282,487	$847,462
38	Raw Water Pumping	1	200	4	3	$135,079	$405,238
39	Gravity Sludge Thickener	20	150	20	1	$234,394	$234,394
41	Sand Drying Beds	4,800	400,000	5333	6	$69,103	$414,620
44	Administration, Laboratory, and Maintenance Building	1	200	10	1	$280,442	$280,442

SUB TOTAL PROCESS COSTS	$25,340,400
YARD PIPING 10%	$2,534,040
SITEWORK LANDSCAPING 5%	$1,267,020
SITE ELECTRICAL & CONTROLS 20%	$5,068,080
TOTAL CONSTRUCTION COST	$34,209,540
ENGINEERING, LEGAL & ADMINISTRATIVE COST 35%	$11,973,339
TOTAL PROJECT COST	$46,182,880

Appendix A9b

Predesign Cost Estimate for a Conventional Treatment Plant with Ozone GAC Filters Figure 2.4.2c with Capacity of 100 MGD

No.	Process	Applicable Range		Quantity Per Unit Process	Number of Process Units	Process Cost Per Unit	Total Process Cost
		Minimum	Maximum				
1b	Chlorine storage and feed 1 -ton cylinder storage	200	10,000	6000	2	$2,039,578	$4,079,155
4	Ozone Generation	10	3,500	2500	2	$6,069,205	$12,138,410
5	Ozone Contact Chamber	1,060	423,000	173750	4	$261,463	$1,045,851
6	Liquid Alum Feed - 50% Solution	2	1,000	400	3	$454,749	$1,364,246
8	Polymer Feed	1	220	200	2	$3,398,249	$6,796,498
12	Sodium Hydroxide Feed - 50% Solution	1	1,600	4000	2	$862,221	$1,724,443
19	Rapid Mix G = 900	800	145,000	17400	1	$193,231	$193,231
21	Flocculator G = 50	0.015	7	0.525	4	$705,106	$2,820,425
25	Gravity Filter Structure	140	28,000	1450	12	$1,843,771	$22,125,254
27	Filtration Media - Dual Media	140	28,000	1450	12	$95,007	$1,140,085
29	Filter Backwash Pumping	90	1,500	600	2	$330,937	$661,875
31	Air Scour Wash	140	27,000	1450	2	$569,143	$1,138,286
32	GAC Filtration/Adsorption Bed	140	28,000	1450	14	$1,843,771	$25,812,796

#	Description						
33	GAC Media	140	28,000	1450	14	$760,056	$10,640,790
34	Wash Water Surge Basin - (Holding Tank)	9,250	476,000	240000	1	$1,608,950	$1,608,950
35	Wash Water Storage Tank - (Waste Holding)	19,800	925,000	250000	2	$261,885	$523,771
36	Clear Water Storage - Below Ground	0.011	8	8.000	2	$6,236,761	$12,473,521
37	Finished Water Pumping - TDH = 100ft - (30.8mts)	1.45	300	31.25	4	$902,691	$3,610,764
38	Raw Water Pumping	1	200	30	4	$525,767	$2,103,069
39	Gravity Sludge Thickener	20	150	10	3	$94,864	$284,592
42	Filter Press	30	6,600	1000	1	$1,802,536	$1,802,536
44	Administration, Laboratory, and Maintenance Building	1	200	100	1	$1,001,943	$1,001,943

SUB TOTAL PROCESS COSTS	$115,090,491
YARD PIPING 10%	$11,509,049
SITEWORK LANDSCAPING 5%	$5,754,525
SITE ELECTRICAL & CONTROLS 20%	$23,018,098
TOTAL CONSTRUCTION COST	$155,372,163
ENGINEERING, LEGAL & ADMINISTRATIVE COST 35%	$54,380,257
TOTAL PROJECT COST	$209,752,420

Appendix B

Additional Cost Indexes,
Charts and Tables

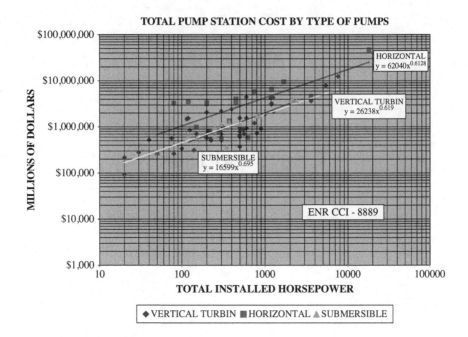

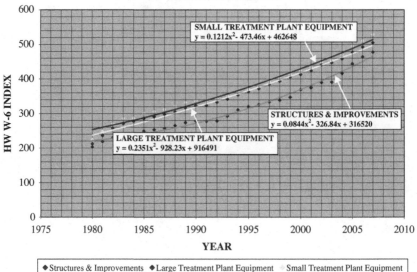

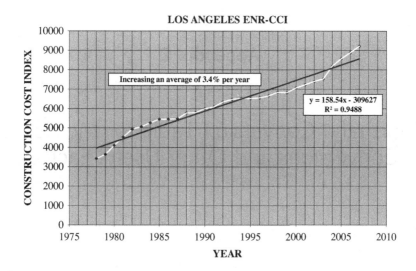

LOS ANGELES ENR-CCI

Increasing an average of 3.4% per year

y = 158.54x - 309627
$R^2 = 0.9488$

Appendix C

Shorthand Conversion Table for Metric (SI) Conversions and Other Useful Factors

Length :
- 1 ml $= 2.54 \times 10^{-3}$ cm
- 1 in $= 2.54$ cm
- 1 ft $= 30.48$ cm
- 1 yd $= 0.914$ m
- 1 mile $= 1.61$ km

Area :
- 1 ft^2 $= 0.093$ m^2
- 1 yd^2 $= 0.836$ m^2
- 1 mile^2 $= 2.59$ km^2
- 1 ha $= 10,000$ m^2
  - $= 2.47$ acres

Volume :
- 1 ft^2 $= 28.32$ L
  - $= 7.48$ gal
- 1 gal $= 3.785$ L
- 1 yd^3 $= 0.7646$ m^3
- 1 acre-ft $= 0.326$ million gallons
  - $= 1,235$ m^3
- 1 m^3 $= 35.31$ ft3
  - $= 264$ gal

Weight :
- 1 dalton $= 1.65 \times 10^{-24}$ grain
- 1 lb $= 453.6$ g
- 1 ton (short) $= 2,000$ lb
  - $= 907.18$ kg

Temperature : $(32°F) \times 5/9 = °C$

Feeding rate :
- 1 mg/L $= 8.34$ lb/million gallons
- 1 ft^3/s $= 448.5$ gpm
  - $= 0.6463$ mgd
- 1 ft^3/min $= 28.32$ L/min
- 1 mgd $= 1.547$ ft^3/s
  - $= 3,785$ m^3/d
  - $= 3.785$ ML/d
- 1 m^3/s $= 22.826$ mgd
- 1 gpm/ft^2 $= 2.5$ m/h
  - $= 60$ m/d

Pressure :
- 1 psi $= 0.0703$ kgf/cm^2
- (lbf/in^2) (kg/cm^2)
- 1 psig $= 14.7$ lb/in^2
  - $= 2.31$ ft H$_2$O
- 1 kPa (N/m^2) $= 0.145$ psi
- 1 MPa $= 145$ psi
- 1 bar $= 1.0197$ kgf/cm^2
  - $= 14.5$ psi

Power :
- 1 hp $= 550$ ft-lb/s
  - $= 42.44$ Btu/min
- 1 hp $= 0.7457$ kW

Gravitational acceleration :
- 1 g $= 32.2$ ft/s^2
  - $= 9.807$ m/s^2

Glossary

AACEi American Association of Cost Engineers, International (AACEi)

Accuracy of the Estimate The accuracy of a predesign cost estimate is taken from the guidelines of the American Association of Cost Engineers, International (AACEi) as a percentage range for estimating purposes.

Allowance for Additional Direct Costs A percentage allowance is intended to cover work items not yet quantified but known to exist in projects of this type and size.

AWWA American Water Works Association

BAT Best available technology

Benefit-cost ratio The ratio of the present worth of estimated project benefits to present worth of estimated project costs. If the ratio is 1.0 or higher (benefits > costs), the project is considered worthwhile; it does not mean that the project should be built—there are many projects and limited resources. SOURCE: "Glossary for Cost and Risk Management," Washington State Department of Transportation (WSDOT), March 2007

CMS Construction management services

CSI Construction Specification Institute

Construction Administration Costs The Base Costs of administration, management, reporting, design services in construction and community outreach, and so on that are required in the Construction Phase.

Construction Contingency A markup applied to account for substantial uncertainties in quantities, unit costs and the possibility of currently unforeseen risk events related to quantities, work elements or other project requirements. SOURCE: "Glossary for Cost and Risk Management," WSDOT

Construction Costs Construction costs are the sum of all individual items submitted in the successful contractor's winning bid through

191

progress of the work, culminating in the completed project, including change order costs.

Construction Cost Trending   The preparation and updating of the project construction cost estimate over time. As the design process continues, the project becomes more defined, and as more detailed engineering data becomes available, "trending" provides a basis for the analysis of the effects of these changes.

Contingencies   Contingencies are defined as specific provisions for unforeseeable cost elements within the defined project scope. SOURCE: AACE International Recommended Practice No.18R-97, AACEi, 1997

Cost Indexes   Cost indexes are a measure of the average change in price levels over time, for a fixed market basket of goods and services.

Design Allowance   See Allowance.

DAF   Dissolved air flotation

DHS   Department of Health Services

ENR   Engineering News Record

ENR CCI   Engineering News Record Construction Cost Index

EPA   U.S. Environmental Protection Agency

Escalation   The total annual rate of increase in cost of the work or its sub-elements. The escalation rate includes the effects of inflation plus market conditions and other similar factors. See also inflation. SOURCE: "Glossary for Cost and Risk Management," WSDOT

Estimate   A quantitative assessment of the likely amount or outcome. Usually applied to project costs, resources, effort, and durations and usually preceded by a modifier (i.e., preliminary, conceptual, order-of-magnitude, etc.).

ESWTR   Enhanced Surface Water Treatment Rule

ft   Foot or feet

ft^3/h   Cubic feet per hour

ft^3/min   Cubic feet per minute

ft^3/s   Cubic feet per second

Forecasts   Estimates or predictions of conditions and events in the project's future based on information and knowledge available at the time of the forecast. Forecasts are updated and reissued based on work performance information provided as the project is executed. SOURCE: Project Management Body of Knowledge *(PMBOK)*, Third Edition

Future Costs   Costs that are escalated by projected inflation rates to specific points in time, consistent with a particular project schedule. SOURCE: *PMBOK*, Third Edition

g   Acceleration of gravity

g   Gram

GAC   Granular activated carbon

gpcd   Gallons per capita per day SOURCE: World Health Organization, *Water a Shared Responsibility*, The United Nations World Water Development Report 2, 2006

gph   Gallons per hour

gpm   Gallons per minute

gpm/ft^2   Gallons per square foot

HVAC   Heating, ventilation, and air conditioning

Historical Information   Documents and data on prior projects, including project files, records, correspondence, closed contracts, and closed projects.

Inflation   The increase in the price of some set of goods and services in a given economy over a period of time. It is measured as the percentage rate of change of a cost index. Inflation's cause is thought to be too much money chasing too few goods.

lb   Pound

Market Conditions   Market conditions are the consequence of supply-and-demand factors, which determine prices and quantities in a market economy and which are separate from inflation.

MCL   Maximum contaminant level

MDG   Millennium Development Goals SOURCE: World Health Organization in *Water a Shared Responsibility,* The United Nations World Water Development Report 2, 2006

MF   Microfiltration

mg   Milligram

MGD   Million gallons per day

mL   Milliliter

NIPDWR   National Interim Primary Drinking Water Regulations

NPDES   National Pollutant Discharge Elimination System

NPDWR   National Primary Drinking Water Regulations

NIPDWR   National Interim Primary Drinking Water Regulations

O&M   Operation and maintenance

PAC   Powdered activated carbon

Parametric Estimating   An estimating technique that uses a statistical relationship between historical data and other variables (e.g., lane miles, square footage, etc.) to calculate an estimate for activity parameters such as scope, cost, budget, and duration. Accuracy is dependent on the sophistication and the underlying data built into the model. An example for the cost parameter is multiplying the planned quantity of work to be performed by the historical cost per unit to obtain the estimated cost. SOURCE: *PMBOK*, Third Edition

PMBOK   An acronym meaning Project Management Body of Knowledge. The term PMBOK^{TM} is used by the Project Management Institute to refer to their *Guide to the Project Management Body of Knowledge* publication.

PMI   Project Management Institute, Inc.

ppd   Pounds per day

ppm   Parts per million

psi   Pounds per square inch

Probability   An estimate of the likelihood that a particular risk event will occur, usually expressed on a scale of 0 to 1 or 0 to 100 percent. In a project context, estimates of probability are often subjective, as the combination of tasks, people, and circumstances is usually unique. SOURCE: "Glossary for Cost and Risk Management," WSDOT

Range   The difference between the upper and lower values of a set of numbers or results, either measured absolutely or related to confidence levels. SOURCE: "Glossary for Cost and Risk Management," WSDOT

Range Cost Estimate A cost estimate that gives a range of costs, related to specific confidence levels. SOURCE: "Glossary for Cost and Risk Management," WSDOT

RCML Recommended Maximum Contaminant Level

Risk The combination of the probability of an uncertain event and its consequences. A positive consequence presents an *opportunity*; a negative consequence poses a *threat*. SOURCE: "Glossary for Cost and Risk Management," WSDOT

RO Reverse osmosis

SDWA Safe Drinking Water Act

SMCLs Secondary Maximum Contaminant Levels

SWTR Surface Water Treatment Rule

UBC Uniform Building Code

UF Ultra-filtration

Ultimate Cost Actual cost at completion of all work elements, including all outside costs, changes, and resolution of risk and opportunity events.

Uncertainty The lack of complete knowledge of any outcome. Economist Frank Knight (1921) *Risk, Uncertainty, and Profit,* University of Chicago established the important distinction between risk and uncertainty: "Uncertainty must be taken in a sense radically distinct from the familiar notion of Risk, from which it has never been properly separated. . . . The essential fact is that 'risk' means in some cases a quantity susceptible of measurement, while at other times it is something distinctly not of this character; and there are far-reaching and crucial differences in the bearings of the phenomena depending on which of the two is really present and operating. . . . It will appear that a measurable uncertainty, or 'risk' proper, as we shall use the term, is so far different from an unmeasurable one that it is not in effect an uncertainty at all." SOURCE: Wikipedia

WHO World Health Organization

Bibliography

AACE, "Cost Estimate Classification System – as Applied in Engineering," Procurement, and Construction for the Process Industries, AACE International Recommended Practice No. 18R-97, AACEi, 1997.

AWWA and ASCE, *AWWA and ASCE Water Treatment Plant Design*, Third Edition New York: McGraw-Hill, 1997.

California Public Utilities Commission, "Adjusting and Estimating Operating Expenses of Water Utilities," (Exclusive of Taxes and Depreciation), California Public Utilities Commission, Water Division, Standard Practice No. U-26, San Francisco, California, July 2002.

EPA, *Construction Costs for Municipal Wastewater Treatment Plants: 1973–1978*, April 1980 USEPA/430/9-80-003 FRD-11.

EPA, *Estimating Costs for Water Treatment as a Function of Size and Treatment Plant Efficiency*, August 1978 USEPA/600/2-78-182.

EPA, *Innovative and Alternative Technology Assessment Manual,* February 1980 USEPA/430/9-78-009 MCD-53.

EPA, "National primary Drinking Water Regulations: Interim Enhanced Surface Water Treatment: Final Rule," Federal Register, 40 CFR Parts 9, 141-142, (December 16, 1998).

EPA, *Operation and Maintenance Costs for Municipal Wastewater Facilities*, September 1981 USEPA/430/9-81-004 FRD-22.

EPA, *Treatability Manual, Volume VI, Cost Estimating*, July 1980 USEPA/600/8-800-042d.

Kawarmura, Susumu, *Integrated Design and operation of Water Treatment Facilities,* Second Edition , Hoboken, NJ:: Wiley, 2000.

Merit, F.S., ed., *Standard Handbook for Civil Engineers*, Third Edition, New York: McGraw-Hill, 1983.

Montgomery, J. M., Consulting Engineers, *Water Treatment: Principles and Design,* Hoboken, NJ: Wiley 1985.

PMI, Inc. *Project Management Body of Knowledge*, Third Edition, Newtown Square, PA:: PMI, Inc., 2004.

"Uncertainty," Wikipedia, the free encyclopedia (http://en.wikipedia.org/wiki/Uncertainty), 2007

Washington Stat Department of Transportation (WSDOT), "Glossary for Cost and Risk Management," March 2007

World Health Organization. World Health Organization in Water a Shared Responsibility, The United Nations World Water Development Report 2, 2006

Index